Praise for *How to Read Numbers*

'A vital plea to take statistics more seriously – the prose being as clear and elegant as the numbers'
Sathnam Sanghera, author of *Empireland*

'A charming, practical and insightful guide. You might not even notice how much you're learning – you'll be too busy having fun'
Tim Harford, author of *How to Make the World Add Up*

'Reading this book is strongly correlated with not looking stupid. Highly recommended'   Helen Lewis, author of *Difficult Women*

'An excellent guide to everyday statistics . . . The authors do a splendid job of stringing words together so smartly that even difficult concepts are explained and so understood with ease. [A] timely and lively book'   Manjit Kumar, *The Times*

'Wonderfully written – incredibly readable. It should be made compulsory reading for everyone before they leave school'
Evan Davis

'An erudite, enlightening guide to the numbers we read in the news – and why they are so often wrong. The authors make sense of dense material and offer engrossing insights into sampling bias, statistical significance and the dangers of believing the casual language used in newspapers'   *Independent*

'[A] fascinating, easy-to-read explanation of how to interpret numbers in the news . . . Their enlightening book provides us with the tools to spot when we're being led astray'
Nick Rennison, *Daily Mail*

'A great com¹ ˙   ⁻   ble'
Mishal Husain

Also by Tom Chivers

The Rationalist's Guide to the Galaxy

How to Read Numbers

A Guide to Statistics in the News
(and Knowing When to Trust Them)

TOM CHIVERS
AND DAVID CHIVERS

WEIDENFELD & NICOLSON

First published in Great Britain in 2021 by Weidenfeld & Nicolson
This paperback edition first published in Great Britain in 2022
by Weidenfeld & Nicolson,
an imprint of The Orion Publishing Group Ltd
Carmelite House, 50 Victoria Embankment
London EC4Y 0DZ

An Hachette UK Company

11 13 15 17 19 20 18 16 14 12

A CIP catalogue record for this book is
available from the British Library.

ISBN (Mass Market Paperback) 978 1 4746 1997 4
ISBN (eBook) 978 1 4746 1998 1
ISBN (Audio) 978 1 4746 2088 8

Typeset by Input Data Services Ltd, Somerset

Printed and bound in Great Britain by Clays Ltd, Elcograf S.p.A.

www.weidenfeldandnicolson.co.uk
www.orionbooks.co.uk

Dedicated to our grandparents, Jean and Peter Chivers

Contents

Introduction   1

Chapter 1: How Numbers Can Mislead   7

Chapter 2: Anecdotal Evidence   15

Chapter 3: Sample Sizes   21

Chapter 4: Biased Samples   29

Chapter 5: Statistical Significance   35

Chapter 6: Effect Size   43

Chapter 7: Confounders   47

Chapter 8: Causality   55

Chapter 9: Is That a Big Number?   63

Chapter 10: Bayes' Theorem   69

Chapter 11: Absolute vs Relative Risk   77

Chapter 12: Has What We're Measuring Changed?   83

Chapter 13: Rankings   91

Chapter 14: Is It Representative of the Literature?   97

Chapter 15: Demand for Novelty   103

Chapter 16: Cherry-picking   113

Chapter 17: Forecasting 119

Chapter 18: Assumptions in Models 127

Chapter 19: Texas Sharpshooter Fallacy 133

Chapter 20: Survivorship Bias 141

Chapter 21: Collider Bias 149

Chapter 22: Goodhart's Law 157

Conclusion and Statistical Style Guide 163

Acknowledgements 173

Notes 175

Index 195

Introduction

Numbers do not feel. Do not bleed or weep or hope. They do not know bravery or sacrifice. Love and allegiance. At the very apex of callousness, you will find only ones and zeros.
Amie Kaufman, *Illuminae*

Numbers are cold and unfeeling. People often dislike them for that reason, and it's easy to understand why. At the time of writing, newspapers still report on daily death tolls from Covid-19, the pandemic which began to sweep around the world in the first half of 2020. When, in Britain, those daily tolls dropped down to the mere hundreds, where before they had been in their thousands, it felt like a light at the end of the tunnel.

Every one of those people, though, was an individual; they were unique. We can talk about the number of people who died during the pandemic – 41,369 in Britain by August, or 28,646 in Spain; or however many will have died around the world once the disease has eventually run its course, if it ever does. But that stark number tells us nothing about those individuals. They all had stories – who they were, what they did, who they loved and who loved them back. They will have been mourned.

Representing all those lost lives with a simple number – 'today X people died' – seems both harsh and stark; it ignores all the grief and heartbreak. It elides all that individuality, all those stories.

But if we hadn't recorded daily death rates, and therefore kept track of the spread of the disease, it is very likely that many more people would have died. Many more unique, individual stories would have been brought to premature ends. We just wouldn't have known how many.

In this book, we're going to talk a lot about numbers: about how they're used in the media, and about how they can go wrong – and give misleading impressions. But along the way we will need to remind ourselves that those numbers stand for something. Often they will represent people, or if not people, then things that matter to people.

This is, sort of, a book about maths. You may think that you are bad at maths and you may be worried you won't understand it. You are not alone. Almost everyone seems to think they are bad at maths.

David teaches economics at the University of Durham. His students need to get an A in A-level maths to be admitted to the course, but quite a lot of them still say they're bad at maths. Tom thinks he is pretty bad at maths, but he has won two awards from the Royal Statistical Society for 'statistical excellence in journalism' (he likes to drop that into conversation from time to time). David, too, sometimes thinks he's bad at maths, and he *literally teaches maths* to people who are, themselves, good at maths.

You are probably better at maths than you think too. What you might not be particularly good at is mental arithmetic. When we think of people who are 'good at maths', we tend to think of people like Carol Vorderman or Rachel Riley off *Countdown*, people who can quickly do sums in their heads. They *are* good at maths, of course; but if you can't do those sums in your head, it doesn't mean that you're not good at it.

Most of the time, we think of maths as having a right answer and a wrong answer. Again, that's not really the case a lot of the time, at least in the sort of maths we're talking about. For instance, take an apparently simple, if horrifying, number: the total death count from Covid-19. What number should we use?

Should we talk about 'confirmed' deaths, where the diagnosis was established with a test? Or should we talk about 'excess' deaths, comparing the number of people who have died this year to the statistical average from the last few years? The two will give you very different answers, and which one we should use depends on what question we're trying to answer. Neither is wrong; but neither is the 'correct' answer either.

What's important is understanding why these numbers aren't clear-cut, and why sometimes what sounds straightforward is in fact more complicated – especially since it is easy to use numbers to mislead or obfuscate, as people (notably but not exclusively politicians) have a tendency to do. These debates affect our lives, and our ability to participate in democracy. By analogy, it's hard to have a functioning democratic state without a literate population; we need to be able to understand the policies our leaders are putting in place, in order to vote knowledgeably for or against those leaders when it comes to election time.

But it's not enough to only be able to understand words. You also need to have some grasp of numbers. Our news increasingly comes in number form: police-reported crimes go up and down; a nation's economy shrinks or grows; the latest figures on deaths and cases from Covid-19 are released. In order to understand the world around us, we may not need to be good at maths, but we do need to understand how numbers are made, how they're used and how they can go wrong, because otherwise we'll make bad decisions, as individuals and as a society.

Sometimes, it's fairly clear how misunderstanding the statistics could lead to bad decisions: if we don't know how many people have Covid-19, for instance, then we can't judge the appropriate response. In others – such as those we'll discuss elsewhere in this book, cases like whether or not bacon causes cancer, or whether or not drinking fizzy drinks makes you violent – it might not be so obvious. But we all use these numbers, consciously or other-wise, to help us navigate the world. Drinking red wine, taking exercise, investing money – we do these things on the basis that we think their benefits (to pleasure, health or wealth) outweigh their risks. We need to know what those benefits and risks are, and how big they are, if we're going to make those decisions wisely. Often, we're getting our understanding of those benefits and risks from the news.

You can't rely on news organisations to give you those numbers straight, without exaggeration or cherry-picking. That's not necessarily because they're trying to deceive you; it's just because they are trying to report exciting, interesting or shocking things, so that you buy their papers or watch their shows. It's also because they – and we – crave narrative: stories in which problems have identifiable causes and solutions. And if you're selecting numbers by how exciting, interesting or shocking they are, you're likely to pick quite a lot that are wrong or misleading.

Also, while journalists are usually clever and (despite the stereotype) well intentioned, they're not traditionally very good with numbers. That means the numbers you read in the news tend to be wrong. Not always, but often enough that it is wise to be wary.

Fortunately, the ways numbers get misrepresented are often predictable: for instance, they can be cherry-picked, by taking an outlier or using a particular starting point, or by chopping up the data repeatedly until you find something; they can be exaggerated, by using a percentage increase rather than the absolute change; they can be used to suggest causation, when really it's just a correlation; and many other ways. This book will arm you with the tools you need to spot a few of them.

We don't want to suggest that you can't ever trust any number you read. We just want to help you make better decisions about which ones to trust, and when.

We've tried to keep the maths to a minimum. Almost everything that looks like an equation has been cut out and put into boxes outside the main text; you can read them if you like, but if you don't it won't limit your understanding.

Occasionally we haven't been able to avoid some

technical concepts, so you will come across things like 'p=0.049' or 'r=-0.4'. Don't worry. These are just shorthand for some fairly simple, real-life, concrete ideas which we are confident you'll be able to grasp.

We've divided the book into twenty-two short chapters. Each one looks at a way in which numbers can mislead, using examples taken from the news. By the end of each chapter, we hope you will understand the problem and know how to spot it in future. We think it's probably best to read the first eight chapters first – they contain some of the things that will help you understand the rest – but if you want to dip in and out, that's also fine; if we refer to a concept we've discussed before, we'll flag it up.

At the end of the book we make a few suggestions as to how the media could do it better – how some of the mistakes that we discuss can be avoided. We like to think of it as a statistical style guide, and we'd love it if you joined us in encouraging the media outlets that you watch and read to start using it.

So let's get to it.

Chapter 1

How Numbers Can Mislead

While it is easy to lie with statistics,
it is even easier to lie without them.
Attributed to the statistician Frederick Mosteller

Covid-19 has provided the world with a high-stakes, high-speed lesson in statistical concepts. The population suddenly found itself having to understand exponential curves, infection fatality rates vs case fatality rates, false positives and negatives, uncertainty intervals. Some of those were obviously complex – but even the ones that felt like they should have been simple, like the number of people who died from the virus, turned out to be slippery. In this first chapter, we'll have a look at how an apparently straightforward number can mislead in surprising ways.

One number that we all had to come to grips with early on was R. It is highly unlikely that in December 2019 more than one person in every fifty would have known what the R value was, yet by the end of March 2020 it was being discussed almost without explanation on mainstream news broadcasts. But because numbers can misbehave in subtle ways, well-intentioned efforts to inform readers of how R changed ended up misleading people.

Here's a reminder: R is the *reproductive number* of something. It can apply to anything that spreads or reproduces – internet memes, humans, yawns, new technologies. In infectious-disease epidemiology, it's how many people, on average, will be infected by a single person with the disease. If a disease has an R of five, on average each infected person will infect five other people.

Obviously, it's not as simple as that; it's an average. An R value of five could mean that, if you had 100 people, every single one infects exactly five people; or it could mean that ninety-nine of them infect nobody at all and one of them infects 500 people. Or anything in between.

It's also not a constant. The R of a new disease at the very start of an outbreak, when no one in the population has immunity to the pathogen, and no countermeasures are likely to be in place – such as social distancing or mask-wearing – might be very different to later on. One goal of public-health policy during an outbreak is to lower R, with vaccinations or behaviour changes, because if R is greater than one the disease will spread exponentially, and if it's below one it will dwindle away.

But, considering these complications, you'd think that there would, in the case of a virus, be one simple rule: if R gets higher, that is bad. So you were probably unsurprised by the tone of the headlines in the British press in early May 2020, warning that 'the virus's R rate may have gone back UP'[1] due to a 'spike in care home infections'.[2]

But, as with everything else, it's a bit more complicated than that.

Between 2000 and 2013, the US median wage went up by about 1 per cent, in real terms (i.e., adjusted for inflation).[3]

You do not need to read this box, but if you can't remember the difference between a median and a mean average, go ahead.

You might remember 'mean', 'median' and 'mode' averages from school. The 'mean' is the one you probably know: it's what you get if you add all the values together and then divide them by the number of values. The 'median' is the middle value in a series.

Here's the difference. Imagine there are seven people in your population, and one of them earns £1 a year, one of them £2, one of them £3, and so on up to £7. If you add all those values together, you get (1+2+3+4+5+6+7) = 28. You divide 28 by the number of people, seven, and you get £4. So your *mean* average is £4.

To get the *median*, instead of adding them together, you line them all up in a row – so the person earning £1 on the far left, the person earning £2 next, and so on, with the person earning £7 standing on the right. Then you see who's in the middle. In this case, it's the person earning £4. So your *median* average is also £4.

Now imagine that the person who earns £7 sells her tech start-up to Facebook for a billion pounds. Suddenly, your mean average is (1+2+3+4+5+6+1,000,000,000)/7 = £142,857,146. So even though six out of the seven people are in the same situation they were before, the 'average person' in the group (going by the mean, at least) is a multimillionaire.

In unevenly distributed situations like this, statisticians often prefer to use the median. If we do that, we line up our people from left to right again, and the person in the middle is still the person earning £4. In a real population of millions of people, this will tell you more about what the population is like than the mean will, especially if the mean is distorted by a few ultra-high-earners at the upper end of the income distribution.

The *mode*, meanwhile, is just the most common value. So if you have seventeen people earning £1, twenty-five people earning £2, and forty-two people earning £3, then the *modal* average is £3. It gets a bit more complicated when statisticians use it to describe continuous quantities, like height, but let's just forget about that for now . . .

The median wage going up sounds like a good thing. But when you look at the population in smaller groups, you notice something strange. The median wage for people who hadn't completed high school had gone down, by 7.9 per cent. The median wage for high-school graduates had gone down, by 4.7 per cent. The median wage for people who'd attended university but not got a degree had gone down, by 7.6 per cent. And the median wage for people who'd achieved a degree had gone down, by 1.2 per cent.

People who did complete high school and people who didn't; people who did complete college and people who didn't. So the median wage of *every single educational group* went down, and yet the median wage of the population as a whole went up.

So what's going on?

What's going on is that even though the median wage of people with degrees went down, the number of people *with* degrees went up considerably. As a result, the median starts doing strange things. This is called Simpson's paradox, after the British code-breaker and statistician Edward H. Simpson, who described the phenomenon happening here in 1951.[4] It doesn't just apply to medians – it can happen with means as well – but in our example we will use medians.

Let's imagine that there are eleven people in the population. Three of them dropped out of high school and earn £5 a year; three of them completed high school and earn £10 a year; three of them dropped out of university and earn £15 a year; and two of them got a bachelor's degree and earn £20 a year. The median wage for the population as a whole (that is, the wage of the middle person in the distribution: see box on previous page) is £10.

Then, one year, the government makes a huge drive to get more people through high school and university. But at the same time, the average wage of each group falls by £1. Suddenly there are two high-school dropouts earning £4, two high-school graduates earning £9, two university dropouts earning £14, and five university graduates earning £19. For every group, the median wage has gone down; but for the group as a whole the median wage has gone up, from £10 to £14. Something like this, but with bigger numbers, happened in the real US economy between 2000 and 2013.

SIMPSON'S PARADOX

NO HIGH SCHOOL			HIGH SCHOOL			DIDN'T FINISH UNI			DEGREE	
£5	£5	£5	£10	£10	**£10**	£15	£15	£15	£20	£20

NO HIGH SCHOOL		HIGH SCHOOL		DIDN'T FINISH UNI		DEGREE				
£4	£4	£9	£9	£14	**£14**	£19	£19	£19	£19	£19

It's surprisingly common. For instance, black people in the USA are more likely to smoke than white people; but when you control for education, you see that *in every educational subgroup*, black people are *less* likely to smoke. It's just that a lower proportion of black people are in the higher educational subgroups, which tend to smoke less.[5]

Or here's a famous example. In September 1973, 8,000 men and 4,000 women applied to grad school at the University of California in Berkeley. Of those, 44 per cent of the men were admitted and just 35 per cent of the women.

But if you looked at the data more closely, you'd notice that at almost every department of the university, female applicants were *more* likely to be granted admission. The most popular department admitted 82 per cent of women who applied and just 62 per cent of men; the second most popular admitted 68 per cent of women and 65 per cent of men.

What was happening was that women were applying for much more competitive departments. One department had 933 applicants, of whom 108 were women. It admitted 82 per cent of those women, and 62 per cent of the men.

Meanwhile, at the sixth most popular department, there were 714 applicants, of whom 341 were women. That department admitted just 7 per cent of the women, and 6 per cent of the men.

But if you combine the data from the two departments together, there were 449 women who applied and 1,199 men. Of the women, 111 were admitted, or 25 per cent; of the men, 533, or 44 per cent.

Once again, in both departments individually, women were more likely to be admitted; but in both departments together, women were *less* likely.

What's the best way of looking at it? Well, it depends. You could argue that in the case of US wages, the population median is more informative, because the median US person's wage has gone up (because more Americans now finish college and high school). Or you could argue that the average woman is more likely to be admitted than a man, whatever her choice of department. But, equally, you could point out that for those people who don't have a high-school diploma, the situation has got worse; and you could point out that the departments that women want to apply to are apparently under-resourced, because they can only accept a tiny fraction of the people who want to go there. The trouble is that, in Simpson's-paradox situations, you can use the same data to tell diametrically opposed stories, depending on what political point you want to make. The honest thing to do is to explain that the paradox is present.

Let's get back to the R value of Covid-19. It went up, so the virus is spreading to more people, which is bad.

Except, of course, it was more complicated than that. There were two quasi-separate 'epidemics' going on at the same time – the disease was spreading differently in care homes and hospitals to how it was spreading in the wider community.

We don't know the real numbers, because they weren't released in that much detail. But we could do a similar sort of thought experiment to the one above. Imagine that there were 100 people with the disease in care homes, and another 100 in the community. On average, each person in the community passes it to two people, and each person in the care homes passes it to three. The R (the average number of people each disease-carrier infects) is 2.5.

Then we go into lockdown. The number of infected people goes down, and so does the R. But – crucially – it does so more in the community than in care homes. Now there are ninety infected people in care homes, who each pass it on to an average of 2.9 people, and there are ten infected people in the community, who each pass it on to an average of one person. The R is now 2.71.* It's gone up! But in both actual groups, the R has gone down.

What's the correct way of looking at it? Again, it's not necessarily obvious; it may be that the overall R is what you care about, since the two epidemics aren't really separate. But it's certainly more complicated than 'the R value going up is bad'.

Simpson's paradox is one example of a wider problem, known as 'the ecological fallacy', that you get when you try to learn about individuals or subgroups by looking at the average of a group. The ecological fallacy is more common than you might think. For readers, and journalists, it's important to understand that the headline figure might conceal a more complex reality, and that, in order to understand it, you might need to break it down further.

---

* That is: $[(90 \times 2.9) + (10 \times 1)]/100 = 2.71$

Chapter 2

Anecdotal Evidence

In 2019, both the *Daily Mail*[1] and the *Mirror*[2] reported on a woman who was told she had terminal cancer, but who underwent alternative therapies –'including hyperbaric oxygen therapy, full body hypothermia, infrared lamp therapy, pulsed electro-magnetic field therapy, coffee enemas, saunas and vitamin C IV therapy' – at a clinic in Mexico, and whose tumour shrank dramatically.

We would assume most readers of this book will have a healthy suspicion of stories like this. But it's an important starting place for understanding how numbers go wrong. It may not seem as though it includes any numbers at all, but it does. The number they're using is hidden, but it's there – the number one. A single person's story, being used to support a claim. It's an example of what we call 'anecdotal evidence'.

Anecdotal evidence has a bad reputation, but it's not inherently wrong. How do we normally decide whether something is true or not? In a very basic way: we check for ourselves, or we listen to other people who have checked.

If we touch a hot pan, and it burns us, we are confident – even with just that one piece of evidence – that hot stoves burn us, that they will always burn us, and that touching them is usually a bad idea. More than that: if someone else tells us that a pan is hot and will burn us, we are usually happy to trust them. We trust other people's experience. In this case, we don't need to do any sort of statistical analysis.

Almost all of the time, as we navigate through life, this approach to evidence is effective. Learning by anecdote or example – by a single person observing a thing and drawing a conclusion – is, very often, all we need. But why? Why is anecdotal evidence OK here, but misleading elsewhere?

It's because, in the case of hot stoves, the outcome of touching it a second time will almost certainly be the same. You could touch the pan over and over again and be pretty sure that it will always burn you. You can never prove it with absolute certainty – perhaps, on the 15,363,205th time, it will feel cold. Or the 25,226,968,547th. You'd need to keep touching the stove until the end of time to be sure it will always burn you, which is probably a bad idea. But most people would probably be happy to assume that if a hot pan burns once, it'll probably burn every time.

There are other things that generally work the same way every time. If you drop something heavy, it always falls down. As long as you remain on Earth, that will happen in a consistent way. The way it happens the first time is a fair example of how it will happen every time. In statistical language, it is *representative of the distribution of events.*

It is hard to avoid using anecdotes. Throughout this book, we'll be using anecdotes – we'll use specific examples of how numbers go wrong in the media, and we hope you'll trust us when we say that they're reasonably representative of how they go wrong more generally.

The problem comes when you start using anecdotes in situations where things are more unpredictable – where the distribution of events isn't straightforward. For instance, imagine that instead of touching a stove, you pet a dog and it bites you. It might be reasonable to conclude that you ought to be more careful, but you can't conclude 'every time you touch a dog, it bites you'. Or instead of dropping a heavy weight, you release a helium balloon. You notice it drifts up and to the west on the wind. You can't conclude 'every time you release a balloon, it goes westward'. The difficulty is in telling which situations are consistent – which are predictable, like the hot stove or the dropped stone, and which are less so, like the helium balloon.

This is a problem in situations such as medicine. You might have some condition – let's say a headache – and you take medicine for it; let's say paracetamol, also known as acetaminophen. For many people, that will be effective. But for a significant percentage of people, it might not be. Each of those people can tell a story, an anecdote, about the drug not working. But on average, it reduces pain. No single anecdote, or even several, will give you the full picture.

The media, though, is built around stories. For instance: 'I've cured my chronic back pain with £19 patch – but NHS won't prescribe it,' says Gary from Essex, as reported by the *Mirror* in 2019.[3] He'd been suffering for years with lower-back pain caused by something called 'degenerative disc disorder', and had had to retire at just fifty-five; he was on a 'phenomenal concoction' of painkillers and anti-inflammatory drugs, worth thousands of pounds a year. And then he started using something called ActiPatch, which 'uses electromagnetic pulses to stimulate the neuromodulation of nerves, which helps to dampen the feeling of pain'. Soon after, he was able to reduce his intake of painkillers by half. Did the patch cure his back pain? Maybe. From the story by itself, we just don't know.

According to a systematic review in the *BMJ* in 2010,[4] about one person in ten worldwide suffers from lower-back pain: millions of people in Britain alone. It's a pretty unpleasant thing to have, and apart from prescribing painkillers and exercise there's not a great deal that doctors can do to treat it, so quite a lot of patients will try alternative methods – perhaps ActiPatch, or something like ActiPatch. And sometimes people get better, all by themselves, whether or not they've used ActiPatch or some other remedy.

So, quite often, a patient will try a new alternative remedy, and then they will get better. And, quite often, the two will be entirely unconnected. So individual anecdotes of people getting better after using some medicine or other can easily be misleading.

The problem is actually worse than we've made it sound, because the media likes news. It seeks out the most interesting and surprising things, or the most heartwarming, or whatever will grab readers' attention. That's not a criticism – there's no way the media could report what happens to the median person every day. But it means that surprising things are more likely than unsurprising things to get into the papers.

To be clear: that may or may not be what happened with Gary and ActiPatch. The fact that the evidence is weak doesn't mean that the conclusion is incorrect; maybe ActiPatch is effective (there is some[5] evidence[6] for similar devices working, and the US FDA cleared ActiPatch for use in treatment of back pain in 2020),[7] and maybe it did work for Gary. It just means that we can't use Gary's story to tell us very much. If we didn't think before that ActiPatch might work, then we shouldn't think so now.

Lower-back pain is a nasty thing to have, and clearly put some hard limits on Gary's life. But on the scale of things, if a lot of people read his story and end up using a patch in the hope that it will reduce their backache, that isn't such a terrible outcome. And perhaps it will do some good – if the treatment works, or if it gives people hope, or reduces pain via the placebo effect – although at a cost, to the health service or patients who pay for it.

Sometimes, it's easy to laugh at. For instance, another story in the *Mail* in 2019[8] claimed that six people recovered from psoriasis after taking homeopathic remedies based on snake venom, whale vomit, decomposed beef and 'the urethral discharge of a man who had gonorrhoea'.

So there may be a tendency to say 'what's the harm?'. In other situations, though – such as the story with which we started this chapter, of the woman who treated her cancer with alternative medicine – it can be more serious. To be clear: there is no good reason to think that hyperbaric oxygen therapy or coffee enemas will treat cancer. But there is good reason to think that there are millions of desperate cancer patients around the world, many of whom will try ever more extreme things in an attempt to get rid of their disease; and also that, sometimes, people get better from cancer. The chance for coincidence, just as with Gary and his back, is enormous.

Perhaps in the case of the woman who treated her cancer with coffee, no harm was done: if her cancer got better, then that's wonderful news, whether or not the coffee helped. And perhaps it gave her hope. But if people are discouraged from using real, evidence-based medicine because they read in the newspaper about someone getting better from using pulsed electro-magnetic field therapy, whatever that is, then that could be dangerous. That is why it is important that we, as a society, understand evidence – how it works, and when it doesn't. That applies to anecdotal evidence, but it applies to all the concepts in this book, when the numbers – and the ways they go wrong – get more complex.

We're not saying that anecdotal evidence is useless. Most of the time we use it very successfully to navigate the world – that restaurant is really nice; you'll like that film; their new album is rubbish. But when it is filtered through the media, the chances of coincidence are high and it becomes largely useless.

In the next chapter, we'll talk about what happens when the numbers get a bit bigger, and why that is a bit better – but only a bit.

Chapter 3

Sample Sizes

Does swearing when you try to lift something heavy make it easier? Apparently, according to a news story in the *Guardian*.[1] It sounds quasi-plausible: most of us will have sworn lavishly while trying to move an IKEA wardrobe that we have unwisely built in the wrong part of the house and now have to carry up the stairs. Maybe it helped.

The story was based on a study carried out at Keele University.[2] In the last chapter, we talked about how news stories based on anecdotal evidence – that is, people's stories about their experience – can easily be misleading. But scientific studies have to be better, right?

Well, sort of.

But not all scientific studies are created equal.

If one person's experience isn't enough to convince you of something, how many should be? It's not a hard-and-fast rule. But let's imagine you want to learn something – say, the height of British men. You're an alien who's never seen a British man, so you have no idea. They might be microns high; they might be the size of stellar clusters. You just don't know.

You could line up every single man in Britain in order of height and measure them. That would give you the full story: that there are only a few very short or very tall people, and that average-height people are the most common. But it sounds like an awful lot of work, even if you threaten them with your Gauss Blaster. Instead, you might take a sample.

A sample is a small piece of something that, you hope,

represents the whole. A free sample of sourdough outside your local boutique bakery tells you something about the loaf; a sample chapter on your Kindle tells you something about the book. A statistical sample does the same thing.

So you go and start sampling your population, by measuring random people on the street. You might be unlucky, and the very first man you measure is 7'. That's given you better-than-nothing information – the 'they're as big as stellar clusters' hypothesis looks much shakier now – but if you conclude that British men are all 7', you'll be badly off the mark. (Another reason why anecdotal evidence is often shaky.)

You're aware of all that, so you keep sampling, and you keep recording heights. You draw a simple graph: each time you find someone who's 5'11", you add a blob to the 5'11" column; each time you find someone who's 6'1", you add a blob to the 6'1" column, and so on.

As you measure more men, you'll notice a shape developing on your chart. It will have lots of marks around the middle, and fewer to either side, forming a sort of humped curve, like an old stone bridge. You'll see the greatest number around the 5'10" mark, and nearly as many falling between 5'8" and 6'1", but you'll see very few at the extremes. That curve will be something like the normal distribution, the famous 'bell curve', centred around the height of the average British man.

That curve will become obvious when you have measured thousands of people, but at first it'll be very bumpy. If you happen to get unlucky and see a few disproportionately tall or short people, it might make your curve look wrong. But assuming you really are sampling the population at random, then on average, the more people you sample, the closer to the population average you will get. (If you're not sampling at random, then you have other problems – see Chapter 4: Biased Samples.)

NORMAL DISTRIBUTION

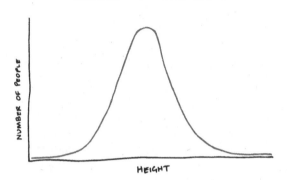

We also need to consider how far people vary from the average. Imagine the average height is 5'10". If everyone is almost exactly average height, with only a few people as tall as 6' or as short as 5'8", then your bell curve would be tall and narrow. If lots of people are 4'10" and lots are 6'10" and every point in between, then the bell curve would be wider and flatter. We can describe how much the data varies like this with a measure called the *variance*.

VARIANCE

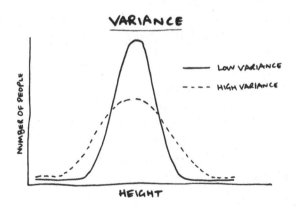

If the variance is low, then you're less likely to see results that are a long way from the average, and vice versa.

You do not need to read or understand this box, but if you would like to know how sample size and the normal distribution work, go ahead.

The casino game craps offers an easy demonstration of how sampling works. It involves simply rolling two dice and adding them up.

There are eleven different results that you can get when you roll two dice and add up the numbers, from 2 to 12. But they're not equally likely.

Imagine you roll one die first, then the other. If you roll a 1 on the first die, then – no matter what you roll on the second – you can't roll a 12 in total. Similarly, if you roll anything *other* than a 1, you can't roll snake-eyes. The number you roll on the first die limits the total you can get, to between X+1 and X+6.

But that also means that *whatever* you roll on the first die, it's still possible to get a 7. If you get a 6, rolling a 1 gets you 7. If you get a 2, then a 5 gets you 7. And so on, up to rolling a 6 and needing a 1. So regardless of the first die, there's a one in six chance of getting a 7.

In total, there are thirty-six different ways the dice can land. Six of them add up to 7, so your chance of rolling a 7 is 6/36, or 1/6. Five of them add up to 8 and five of them add up to 6; four of them add up to 9 and four of them add up to 5; and so on. But there's only one way of getting 2, and only one way of getting 12.

You can prove this mathematically, as we just have, but you can also demonstrate it by rolling them yourself. If you roll the dice thirty-six times, you probably won't see exactly six 7s, exactly five 6s, and so on. But if you roll the dice a million times, you'll see them come up 7 almost exactly one time in six, and snake-eyes one time in thirty-six.

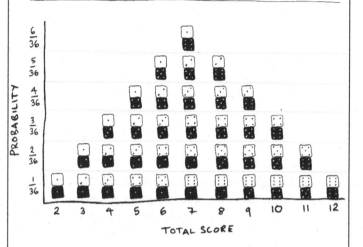

PROBABILITY OF TOTAL SCORE ON TWO DICE

Let's imagine you're trying to find out, empirically, how often you roll 7 on two dice. The basic principle is this: the more times you roll – the larger your sample size – the more confidently you can predict how many 7s you'll see. If you roll twenty times, then there's a 95 per cent chance you'll see between one and six 7s. That's six possible outcomes, more than 25 per cent of your possible range of outcomes.

If you roll 100 times, your 95 per cent probability interval is between eleven and twenty-five: just 15 per cent of the possible outcomes.

If you roll 1,000 times, then there's a 95 per cent chance that you'll roll a 7 between 145 and 190 times. Your range of outcomes has narrowed to just 4.6 per cent of the total.

The same will happen for all the other possible rolls: you'll see snake-eyes turn up closer and closer to exactly 1/36th of the time, and double 6 the same, and all the intervening numbers.

As you 'sample' more and more dice rolls, you get closer and closer to the 'real' distribution.

* As a reward for those of you who stuck with this, it might amuse you to know that Joe Wicks, the gentle soul who guided Britain through lockdown by running PE lessons from his front room via YouTube every weekday, got himself into all sorts of bother with this. He tried to introduce randomness into his workouts by rolling two dice and assigning his exercises a number from 2 to 12, but was then extremely confused when he found he was doing a lot more burpees (assigned to 7) than he was star jumps (assigned to 2).[3] Once he realised his mistake, he introduced a roulette wheel.

With men's height, you've got a simple distribution around the average – again assuming your sampling is random, the more men you measure, the more closely your sample will resemble the population as a whole, just as in the boxed example about dice.

But say you want to work out something else – for example, whether someone who takes a drug gets better faster than someone who doesn't. In this case you're not measuring one thing, you're measuring two: how fast people get better when they take the drug; and how fast they get better when they don't.

You want to find out if the two groups are different. But, just as with height, there's going to be some random variation. If you just picked two people, got one to take the drug and one not to, then you may find that the drug-taking one got better faster. But that might just be because they happened to have a stronger constitution.

So you randomly assign lots of people into two groups, and you give some of them the drug and some of them a placebo. Then you measure the average time it takes both groups to get better, just as you measured the height of the population. You're doing the same thing, essentially – 'sampling' an imaginary 'population' of people who have taken a drug, and another 'population' of people who haven't. If, on average, the people who took the drug get better more quickly, then that suggests that the drug makes you get better more quickly.

The trouble is that, just as with measuring average heights, you might be unlucky. You might accidentally sort all your healthier people into the intervention group, or even just the majority of them. Then it would look as though your drug has made people get better, when in fact it's just that they would have got better more quickly anyway.

Of course, the more people you include in your sample, the less likely it is that those random variations will affect the outcome. The question is, how many do you need to get a good estimate? The answer is: it depends.

It depends on various factors, but one of the most important is the subtlety of the thing you're looking for. The smaller the effect an intervention has, the more people you need to look at to investigate it – in the jargon, the more 'statistical power' you need. This is pretty obvious when you think about it. You don't need a sample size of 10,000 people to investigate the question 'Is being shot in the head bad for you?'

To return to the swearing study: the effect that swearing has on your strength is – you'd expect – probably minimal, if it's real at all. If it was obvious, we'd have noticed, and the Olympic weight-lifting final would have to be broadcast after the watershed.

The swearing study involved two separate experiments, looking at two different measures of strength. One had fifty-two participants and one had twenty-nine. We should note that it was a bit different to the model of study we describe above: some people were asked to try to lift something while swearing, and others were asked to lift something while shouting a non-sweary word, just as in the drug trial we described. But in this case, the two groups were then swapped over, and people who hadn't sworn were asked to, and vice versa. Both groups had their strength measured both times. This is called a 'within-subject' trial design and can reduce the problems of small sample sizes.

As we said, the exact size of the sample you need depends on several things, including the subtlety of the effect you're looking for. And there are statistical tricks that you can do to reduce the chance that you're just seeing a random result.

As a rule of thumb, though, we think you should be wary of any study that has fewer than 100 participants, especially if it's used to make some quite surprising and/or subtle claim; as a study increases in size, all else being equal, your confidence in it should increase. It may be that swearing makes you stronger, but we'd be pretty fucking surprised.

Again, this is all fun and games – who really cares whether swearing makes you stronger? It's an interesting thing if it's true, and it's entertaining, but it is unlikely to be a life-and-death issue.

But that's not true for many things. In the first half of 2020, as the world was flailing around for something – anything – that might treat or prevent Covid-19, scientific papers and preprints (early versions of scientific papers, which have not yet gone through peer review) filled the internet. One of them looked at the use of an antimalarial drug hydroxychloroquine.[4] It was a controlled trial, like the swearing-makes-you-stronger study (although not randomised), and it got sufficient attention that one Donald J. Trump cited it in a tweet.[5] The study found that 'hydroxychloroquine treatment is significantly associated with viral load reduction/disappearance in COVID-19 patients.'

It looked at a total of forty-two patients: twenty-six in the intervention group, who were given hydroxychloroquine; sixteen in the control, who weren't. Even if the study was perfectly well performed in other ways (it wasn't), it would still be vulnerable to the problems of small sample size. Just as it's possible that swearing might in fact make you stronger, it's possible that hydroxychloroquine has some effect on Covid-19. But it's also possible that it doesn't, and it's also possible that it's actively harmful. That study does very little to tell us one way or the other. And yet it made headlines around the world.

Chapter 4

Biased Samples

There was exciting news in the *Sun*[1] and *Daily Mail*[2] in April 2020, revealing that Britain's favourite 'lockdown snack' was (drum roll please): cheese on toast. With 22 per cent of the vote, the dairy-topped hot granary product beat cheese-and-onion crisps into a close, but still disappointing, second, on 21 per cent. Runner-up snacks included bacon sandwiches (19 per cent), chocolate cake (19 per cent) and cheese and crackers (18 per cent).

In the last chapter, we looked at how small sample sizes can throw numbers out, simply by them being randomly wrong. But the snack stories are based on a poll by the online banking company Raisin, which surveyed 2,000 people.[3] So it's probably kosher, right?

Well, there are other ways in which studies can be wrong. The most obvious is that, often, the sample you've taken isn't representative of the population you've taken it from.

In the last chapter we imagined trying to work out the height of the population by measuring people at random. But now imagine that you'd set up your people-measuring stall outside a basketball players' convention. Suddenly you might find that you were seeing a lot of seven-footers strolling past. The average height of your sample jumps up, but the average height of the population is unchanged.

This is called sampling bias. The word 'bias' is usually applied to humans – the referee is biased against my team; the media is biased against my favoured political party. But statistical bias works in much the same way. Imagine you conducted a survey asking: 'What's the greatest football club in English top-flight history?', first on Anfield Road and then on Sir Matt Busby Way. You'd get very different results, because the samples of people you'd get would be very different.

Biased samples are pernicious in a way that small samples aren't. At least with small but random samples, the more data you get, the closer you'll get to the true answer. But with biased samples, getting more data doesn't help and instead can make you more confident in your wrong answer.

For instance, in the run-up to the 2019 UK general election, Jeremy Corbyn – then leader of the Labour Party – and Boris Johnson, the prime minister and Tory leader, held a televised debate.

The political polling company YouGov polled viewers afterward and found that they were evenly split on who had 'won'

the debate, with 48 per cent saying Johnson, 46 per cent Corbyn and 7 per cent saying they didn't know. (Yes, it adds up to 101 per cent. That's what you get sometimes when you round off to the nearest whole number.)

But that caused some debate online. One viral tweet (more than 16,000 retweets as we write) pointed out that other polls had found very different results[4] (see figure on the previous page).

Of the five polls, four found that Mr Corbyn had won the debate handily. The only poll which found otherwise had a fraction of the sample size of any of the others. And yet that one was the only one quoted by the broadcast news channels. Does this show media bias against Mr Corbyn?

It's more likely to be an example of sampling bias. The other four polls were all carried out on Twitter. Twitter polls are usually just harmless fun ('World Cup of Crisps, semi-final: Monster Munch Pickled Onion vs Walkers Cheese & Onion' etc.). But sometimes they're used to ask political questions.

The trouble is that Twitter is not representative of the population. The 17 per cent[5] of the UK population who use Twitter tend (according to a 2017 study)[6] to be younger, more female and more middle-class than the country as a whole. Younger people, women and the middle classes are more likely to vote Labour than the country as a whole. (And, of course, the specific people who saw the tweets in question, and who then responded to them, weren't representative of Twitter as whole.)

It wouldn't help to have asked more people on Twitter. You'd still have the same problem, since you're still polling a non-representative sample. If you polled a million people on Twitter, it would still be polling the population of Twitter, not of the country – you would simply get ever more precise results around the wrong answer.

The trouble is that it's very hard to get a representative sample. If you ask people on Twitter, you don't get people who don't use Twitter. But the same problem is true everywhere. If you run internet polls, you won't get people who don't use the internet. If you ask people on the street, you won't get people who stay at home. Political pollsters used to use landlines, because almost everyone had a landline, and it was a very simple way to randomly sample the population – just dial random numbers. But doing that now would get you a heavily skewed sample, because the sort of people who have landlines (and who answer unknown numbers) are different from those who don't.*

There are ways of sampling the population that help to get around these problems in polling, to some degree, but it will never be perfect; for one thing, you can't force people to do a survey, so you will never be able to fully sample the subset of people who really hate doing surveys. Instead, pollsters do something else: they weight their results.

Imagine you know from census data that the population is 50 per cent male and 50 per cent female. Then you do a poll, trying your best to get a representative sample. Of your 1,000 respondents, you get 400 women and 600 men. You're asking: 'Do you enjoy the TV programme *Grey's Anatomy*?' Your response is that 400 people do like it, and 600 people do not. So you might think that 40 per cent of people like *Grey's Anatomy*. But when you break it down you realise that there's a gender skew: 100 per cent of the women said they like it, compared to just 0 per cent of the men.

---

* Funnily enough, this is a sort of reverse of what happened in the 1936 US election. A telephone poll run by *The Literary Digest* magazine surveyed voters ahead of the contest between Alfred Landon, the Republican governor of Kansas, and Franklin D. Roosevelt. It surveyed 2.4 million voters and predicted a heavy 57–43 per cent win for Landon. The actual result was Landon 38 per cent, Roosevelt 62 per cent. *The Literary Digest* had relied on a phone survey at a time when phones were expensive new technology, owned mainly by the rich, and this had skewed their findings enormously. George Gallup, founder of the polling company Gallup, surveyed just 50,000 people and got a much more accurate result, predicting victory for Roosevelt.

The reason you got 40 per cent is because your sample isn't representative of the population. Luckily, you can easily un-skew it. You simply weight the results: you know the population is 50 per cent women, but your sample is only 40 per cent. You know that 50 is 25 per cent bigger than 40. So you take your result of 400 yeses and add 25 per cent, which gives you 500.

Meanwhile, you do the same for the men. Your sample is 60 per cent men, but you know that in an unbiased sample it would be 50 per cent. You know that 50 is 0.833… times 60, so that gives us a weight of 0.833…

So you take your result of 600 and multiply it by 0.833… and you get 500. Now, your weighted results tell you that exactly 50 per cent of the population enjoy the TV programme *Grey's Anatomy*.

You can do this in more subtle ways. For instance, if you ask people who they voted for at the last election, and you know that 35 per cent of the country voted Labour and 40 per cent Conservative, but 50 per cent of your respondents say they voted Conservative, you can reweight your sample accordingly. Or if you know the age distribution of the population, but your sample returns more old people because you used a landline, then you can reweight for that.

Of course, it relies on you knowing true facts about the population: if you believe that it's 50 per cent men and 50 per cent women but it's actually 60–40, then your weighting will make things worse. But you can often get a grasp of the underlying reality from things like the census, or election results.

There are other ways in which samples can be biased; the most obvious is leading questions. For instance, if you ask people whether we should give medicine to 600 people, their answers will depend on whether you say '200 people will be saved' or '400 people will die', even though those statements are logically identical.[7] This 'framing effect' is visible in polling: for instance, there is a tendency to say 'yes' to yes-no questions (e.g. 'Should the government pay for the treatment?').

*

So, is cheese on toast Britain's favourite snack? Well, perhaps Raisin.co.uk did take great pains to do a representative sample, and perhaps they did reweight their results according to age, sex and voting intention: we don't know. (We have asked! And if they respond, we'll update this bit for reprints, we promise.)

But that sounds like an awful lot of effort to go to for a bit-of-fun poll, so we'd be a little surprised. Most likely they just ran an internet survey and received a disproportionate number of responses from the sort of people who answer internet surveys.

The question is whether the subset of the population which answers internet surveys has the same taste in snacks as the population at large. And perhaps they do. But we simply don't know. All we know is that of the 2,000 people they asked, 22 per cent of them said cheese on toast. That's fine, and interesting in its own right – you can now say things about those 2,000 people. But you probably can't say very much about the population as a whole.

Chapter 5

Statistical Significance

Do men eat more when women are around, to impress them? So claims a headline in the *Daily Telegraph* from 2015.[1] The same study ended up in Reuters[2] and the *Economic Times* of India.[3] The stories claimed that men ate 93 per cent more pizza and 86 per cent more salad if they were with women than if they were with other men; they were based on a study by Brian Wansink, a psychologist at Cornell University's Food and Brand Lab, and two other researchers.[4]

By now, you'll have worked out that the numbers in the stories we mention in this book aren't always entirely trustworthy. But in this case, it's definitely not the journalists' fault. The study in question turned out to be terribly wrong in a way that reveals a lot about how science works and doesn't work. To understand why you can't trust the statistics here, we'll need to go deep into the mechanisms of scientific practice. But if you get through it, a lot of what follows in later chapters will make more sense.

If you've read almost any stories about science or numbers in the news, you'll have come across the term 'statistical significance'. You would be forgiven for thinking that the phrase implies that the statistics that you are reading are significant. Unfortunately, it's a lot more complicated than that. Here's what it means, according to a definition from a 2019 paper:[5]

> Assuming that the null hypothesis is true and the study is repeated an infinite number times by drawing random samples from

the same population(s), less than 5% of these results will be more extreme than the current result.

Does that help? Let's have a go at unpacking it.

Imagine you're trying to find something out. Say, whether reading books called *How to Read Numbers* makes you better at understanding statistics in the news. You could take a lovely big sample of a thousand people. That sample will include some of the many millions of people who have read the book, and also, alas, some who have not. (And let's imagine for the sake of argument that, before anyone read the book, the two groups were the same, even though we know that, in reality, the average buyer of this book will certainly be far cleverer, wiser and more beautiful of face than the rest of the population.)

Next we make everyone in our sample take a simple quiz that measures statistical ability, to see whether people who have read this book do better than those who have not.

Let's suppose that, looking at the data, it seems that people who have read this book did better on the test. How do we know that's not just a fluke? How do we know that they did better because of some real difference, not just random variation? To find out, we could use a statistical technique called *significance testing* (or *hypothesis testing*).

First, we imagine the results we'd expect to see if the book had no effect whatsoever. This is called the 'null hypothesis'. The other possibility – that the book does have some positive effect – is called your 'alternative hypothesis'.

The best way to picture this is with a graph. Under the null hypothesis, we'd expect to see a curve that peaks around the average score, with most people around the middle, but a few people doing really well and a few people doing really badly – like the normal distribution curve in Chapter 3. And we'd expect the average score, and the distribution curve, for the book-readers to be pretty much the same as for the non-readers.

Under the alternative hypothesis, the average score for

book-readers should be higher than for non-readers, and the distribution curve will shift to the right.

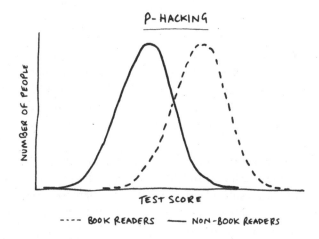

Where it gets complicated is that even under the null hypothesis – that is, even if the book has no effect, and even if, implausibly, the two groups start off exactly equally good at statistics – you'll see some random variation. Some people might just have an off day. One way to imagine it is to think about the film *Sliding Doors*: in one universe Gwyneth Paltrow misses her train and is late for the quiz, so she's flustered and does badly; in another universe she is on time, aces the test, and goes on to fall in love with John Hannah. It probably wouldn't be enough to turn her from dunce to stats genius, but it would be enough to affect her results. There will be *some* level of randomness, however small, in how well everyone does at the test.

If a few non-readers happen to do particularly badly, or a few book-readers happen to do particularly well, it might be enough to change the average noticeably: to make book-readers look like they're doing better than non-readers.

So, let's say that, for whatever reason, your results show that book-readers have done better than non-readers. What you need to do now is see how likely those results (or more extreme results) would be *if the null hypothesis was true* – that is, in our example, if reading the book has no effect, and any variation is just randomness. That's significance testing.

There is no single point at which we can unequivocally say that the null hypothesis is false; in theory, even the most dramatic results could be a total fluke. But the bigger the difference, the more unlikely that fluke is. Scientists measure the chances of coincidence with something called the probability value, or 'p-value'.

The more unlikely something is to happen by random chance, the lower the p-value is: so if there's only a one-in-100 chance that you'd see a result at least that extreme if there was no effect, that would be written as $p=0.01$, or one divided by 100. (What that *doesn't* mean – and this is EXTREMELY IMPORTANT, so important that we will write EXTREMELY IMPORTANT in capitals, twice – is that there is only a one-in-100 chance that the result is wrong. We'll come back to that in a bit, but it is worth flagging here.)

In many parts of science, there's a convention that if p is equal to or less than 0.05 – if you'd expect to see results that extreme no more than 5 per cent of the time – then the finding is 'statistically significant', meaning that you can reject the null hypothesis.

So let's say that when we look at our results, we see that the average score for people who've read our book is indeed higher than the average score for people who haven't. If the p-value of that result is less than 0.05, then we would say that we've achieved statistical significance, and we would reject the null hypothesis (that the book doesn't do anything) in favour of the alternative hypothesis (that the book makes you better at stats). What the p-value is telling us is that, if the null hypothesis were true and we were to run the test 100 times, we would expect to see our book-readers do as well as they have, compared to the non-readers, fewer than five times.

*

Statistical significance is confusing, even for scientists. A 2002 study found that 100 per cent of psychology undergraduates misunderstand significance – as, even more shockingly, did 90 per cent of their lecturers.[6] And another study found that twenty-five of twenty-eight psychology textbooks they looked at contained at least one error in their definition of statistical significance.[7]

So let's clear up some possible misunderstandings. First, it's important to remember that what we call 'statistical significance' is an arbitrary convention. There's nothing magical about p=0.05. You could set it higher, and you'd declare more findings statistically significant; you could set it lower, and you'd declare that more results weren't, and that they could plausibly be flukes. The higher you set it, the greater your risk of a false positive, and the lower you set it, the greater your risk of a false negative. If we are too strict, we might declare that reading our book has had no effect when in fact it did – and, of course, vice versa.

Second, it also doesn't mean that the finding is 'significant' in the usual sense of the word. For instance, if the average score in your non-book-reading group is sixty-five, and the average score in your book-reading group is sixty-eight, that might well be 'statistically significant', but you may not care all that much. 'Statistical significance' is a measure of how likely you are to see something by fluke, not of how important it is.

A third, crucial, point is that it doesn't mean that if you get a finding of p=0.05 there's only a one in twenty chance that your hypothesis is false. That misunderstanding is common and is a big part of why science goes wrong.

The problem is that even though statistical significance at p≤0.05 is completely arbitrary, scientists – and, more importantly, journals – very often treat it as a cut-off point. If your study finds p=0.049, it might get published; if it finds p=0.051, it may well not. Scientists need to get their research published if they want to win grants, achieve tenure, and generally get on in their career. They are heavily incentivised to find statistically significant results.

Let's go back to our book-reading experiment. We really want to show that our book improves statistical ability, so that we get on the *Sunday Times* bestseller list and get to go to all the best cocktail parties. But when we run our experiment, we find that we only get p=0.08.

Well, we think, it could just be bad luck. So we run it again. This time we get 0.11. And we do it again, and again, until eventually we get 0.04. Amazing! We report our findings and live off the book royalties for ever. But this is almost certainly a false positive. If you run an experiment twenty times, then you'd expect to see a one-in-twenty fluke result.

That's not the only way we could do it. We could also chop up our data in lots of different ways. Say that, as well as measuring the score, we measured how quickly people completed the test, or how neat their handwriting was. If we didn't find that book-readers score higher, we could see whether they complete the test faster; if that doesn't find anything, we can see whether their calligraphy has improved. Or you could remove some more extreme results and call them 'outliers'. If we measure enough things, and combine them in enough ways, or make enough small, reasonable-seeming adjustments to the data, then we could guarantee that we'd find *something* just by coincidence.

Let's go back to those stories about men eating more to impress women. In late 2016, Wansink, the lead author, wrote a blog post which would come to sink his career. It was called 'The Grad Student Who Never Said "No"'.[8]

Wansink wrote about a new Turkish PhD student who had arrived in his lab. He had, he said, given her 'a data set of a self-funded, failed study which had null results (it was a one-month study in an all-you-can-eat Italian restaurant buffet where we had charged some people ½ as much as others)'. He told her to go through the data, because 'there's got to be something here we can salvage'.

At his prompting, the PhD student reanalysed the data in dozens of different ways, and – it won't surprise you to learn – found lots of correlations, in exactly the same way as, in our imaginary book-reading study above, we could chop up our data as much as we liked until we found a p<0.05 result. She and Wansink published five different papers from that dataset, including the 'men eat to impress women' study. In it, they found a p-value of 0.02 for men eating more pizza around women, and 0.04 for salad.

But that blog post raised red flags with scientists. Behaviour like this is known as 'p-hacking', massaging the data to get your p-value to a publishable below-0.05 figure. Methodologically savvy researchers started to go through all Wansink's old work, and a source leaked his emails to Stephanie M. Lee, an investigative science journalist at BuzzFeed News. It transpired that he had asked his PhD student to break up the data into 'males, females, lunch goers, dinner goers, people sitting alone, people eating with groups of 2, people eating in groups of 2+, people who order alcohol, people who order soft drinks, people who sit close to buffet, people who sit far away, and so on'.[9]

Other methodological problems were found with Wansink's old papers, and other emails revealed shoddy statistical practice – in one, he suggests that 'we should be able to get much more from this . . . I think it would be good to mine it for significance and a good story'.[10] He wanted the research to 'go virally big time'.

This was a dramatic example. But p-hacking – in less dramatic forms – goes on all the time. It is usually innocent. Academics desperate to get p<0.05, so they can get their paper published, will rerun a trial, or reanalyse the data. You might have heard of the 'replication crisis', in which lots of important findings in psychology and other disciplines have turned out not to exist when other scientists tried to replicate their findings. It was based on a failure of scientists to understand exactly this problem: they kept chopping up their data and rerunning their studies until they found statistically significant results, not realising that by doing so they were rendering their work meaningless. We'll talk more about this in Chapter 15: Demand for Novelty.

Uncovering Wansink's behaviour took months of digging by conscientious, statistically minded researchers and an experienced science journalist. Most of the time, journalists writing about science are writing quick news stories from press releases. They are not going to be able to spot p-hacking even if they have the dataset, which they usually don't. And p-hacked studies have an unfair advantage: because they don't need to be true, it's easier to get them to be exciting. So they tend to turn up in the news a lot.

There's no easy way for readers to spot this in news stories. But it's worth being aware that just because something is 'statistically significant', it doesn't mean that it's *actually* significant, or even that it's true.

Chapter 6

Effect Size

How scared should we be of screen time? Over the last few years there have been all sorts of dramatic claims – notably, that iPhones may have 'destroyed a generation',[1] or that 'for girls, social media use is substantially more harmful than heroin use' (that claim has since been removed from the article).[2] Research into the area is messy and difficult, bogged down by problems of getting good data and avoiding spurious correlations, although the most robust science seems to suggest far less reason to be concerned.[3]

But one area that has got a lot of attention is the link between screens and sleep. One headline in 2014 screamed, rather intemperately, that 'Reading On A Screen Before Bed Might Be Killing You'.[4] It was based on a study in the *Proceedings of the National Academy of Sciences*.[5]

The idea was simple: not getting enough sleep is bad for your health; the study found that reading on lit-up screens reduces how much sleep you get; ergo, the news story reasoned, reading on lit-up screens might be killing you.

First things first. The study does indeed find that how much you use a screen is linked to how much you sleep. Participants were asked to read an e-book before bed on one night, and a normal printed book before bed on another night. (The order they did this in was randomised: some read the printed book first, others the e-book, in case reading one before the other affected the outcomes.)

It found a *statistically significant* result – p<0.01, which you'll remember from Chapter 5 means that if there was no real effect at all, and you ran the study 100 times, you'd expect to see a result as extreme as this less than once. That said, it was a very small study – just twelve subjects – and as we saw in Chapter 3, small sample sizes can lead to strange findings. But sometimes even when studies are small, as long as they're treated with caution, they can be useful in pointing us towards possible avenues of research.

However, as you'll also remember from Chapter 5, 'statistically significant' does not mean 'significant'. If a finding is statistically significant, that just means that there's a good chance that it's real. The other thing you need to consider is the *effect size*. Conveniently, unlike 'statistical significance', 'effect size' means exactly what it says on the tin: the size of the effect.

We're still talking about books, so let's go back to our imaginary experiment from Chapter 5 where we looked at readers of this book. This time, we're going to do it slightly differently. We're going to compare 500 people who read *How to Read Numbers* with 500 who read an inferior book, such as *Middlemarch* or *The Complete Works of William Shakespeare*. And then, instead of measuring the effects on statistical ability, we are going to measure what time they get to sleep and see whether one group is awake later than the other.

When you get the results back, it's clear: all 500 who read *Numbers* fell asleep later than the 500 who read the other.

That would undoubtedly be a statistically significant result. Even without knowing how much of a difference there was, the odds would be astronomical against it being a fluke; odds of one in some number far bigger than the number of atoms in the universe. Assuming the study was well conducted, there is no way that it's not a real effect.

Now imagine that we look at how big the effect is. We see that all 500 people who read *Numbers* did indeed fall asleep later – by precisely one minute.

It's a real effect. It's statistically significant. But it is completely irrelevant to your life. If you're trying to get information that will help you improve your sleep, then this will be of no use to you whatsoever.

Whether something is statistically significant or not can be of huge interest to scientists: if you learn that something correlates with something else, you can investigate that correlation, and you might learn something about the mechanisms behind it. If, for instance, the effect of screen time on sleep is real, then even if it's small it might tell us about how our circadian rhythms work – about whether blue light can help reset our internal body clocks. That could lead to interesting discoveries further down the line. And sometimes even small effects are important: perhaps a cycling team manages to find a way to make a more perfectly round wheel, which knocks one-thousandth of a second off the cyclist's time per mile; it could be enough to make the difference between gold and silver, especially if the team doctor also prescribes them enough asthma medication.

As a reader, though – as someone trying to make sense of the world, and trying to understand how to navigate the risks and difficulties you face in it – whether there's a statistically significant link between two things isn't, of itself, of more than intellectual interest. You might want to use a Kindle instead of a printed book at bedtime so you can switch the lights out and let your partner sleep, for instance. You don't really care about whether it is possible to detect a link. You care about how big the link is.

How big is the effect of reading from a screen before bed? Well: it's tiny. The study's participants were asked to read their books or e-books for *four hours* before bed (*four hours!*). And – as the 'Reading On A Screen Before Bed Might Be Killing You' story didn't quite find space to mention – on the e-book-reading nights, the subjects fell asleep on average ten minutes later. Maybe losing ten minutes' sleep matters, if it happens every night, but who reads for four hours in bed every night anyway?

Interestingly, a later, much larger study looking at young people found much the same thing: there is a correlation between screen use and sleep, but it's small.[6] An extra hour of screen use correlated with between three minutes and eight minutes of lost sleep. That may mask a wide variation – perhaps most children and teenagers are unaffected, but a few are affected very badly. But it doesn't seem as though banning screens before bedtime will help the nation's sleeping habits all that much.

We'd love it if newspapers and the media got used to talking about effect sizes, not just statistical significance. They don't have to go into the technical details, but simply saying 'four hours of reading was associated with about ten minutes of sleep loss' would give readers the information they need to work out whether this is something to care about. And readers should be on the lookout not just for associations – does eating bacon cause cancer? – but how big those associations are (if I eat bacon every day for twenty years, how much more likely am I to get cancer?). If it doesn't seem to be mentioned anywhere in the article, then the likeliest explanation is that the association, the effect, is very small, and the story is less interesting than it sounds.

Chapter 7

Confounders

There's been a lot of controversy over the last few years about vaping. Most anti-smoking and cancer charities think it's a good way to quit smoking, but some people think it's harmful or that it encourages people to take up cigarettes. And in 2019, it was reported that children who smoke e-cigarettes are more likely to use marijuana.[1]

The claim was based on a paper in the journal *JAMA Pediatrics*, which looked at twenty-one other papers and synthesised their results.[2] Papers like this, which aggregate other research, are known as 'meta-analyses'. This meta-analysis found that twelve- to seventeen-year-olds who vaped were about three times as likely to smoke cannabis.

We were just discussing effect sizes, and that sounds like a pretty big one. We'll talk in the next chapter about the difficulties of establishing causality, but it definitely sounds like something worth worrying about here.

But there's something else you need to be aware of when you see that two things – in this case, vaping and marijuana use – are strongly correlated: is there something else that correlates with both? This something else is called a 'confounding variable'.

In case you're not quite following, here's an example. The proportion of deaths linked to obesity worldwide by year correlates with the amount of carbon dioxide emitted by year.[3]

So does carbon dioxide make people fat? Probably not. Instead, what's likely been going on is that the world is getting richer, and as people get richer, they have more money to spend on both high-calorie foods and on carbon-emitting goods like cars and electricity. If you take that into account, the link between carbon emissions and obesity will probably disappear. The third variable, GDP, accounts for most of the link between one and the other.

MULTIPLE VARIABLES

SHARE OF DEATHS FROM OBESITY (%)
AND ANNUAL CO_2 EMISSIONS (GIGATONNES)

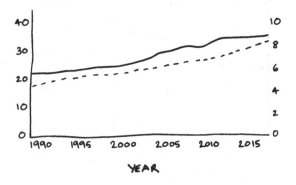

---- SHARE OF DEATHS FROM OBESITY (%)
———— ANNUAL CO_2 EMISSIONS (GIGATONNES)

Another classic example is ice cream and drownings. On days when ice cream sales go up, so do drownings. But obviously ice cream doesn't make people drown. Instead, ice cream sales go up on hot days, because ice cream is nice on a hot day; and so is swimming, which unfortunately leads to some people drowning. Once you adjust for temperature – or 'control' for it, in statistical language – the link goes away. So if you were to look at ice cream sales and drownings just on cold days, or just on hot days, you wouldn't see a link.

This is important when you're discussing effect sizes. It can look like one variable is strongly linked with another – for instance, vaping and marijuana use. But finding out whether this effect is real, or whether it's really caused by some other variable – some 'confounder' – is difficult.

The studies that the vaping meta-analysis looked at did control for possible confounders – age, sex, race, parental education, tobacco-smoking, drug use; it differed by paper. And some papers found stronger links than others. For instance, one – which controlled for sex, race and school grade – found a huge correlation: vapers were about ten times as likely to smoke weed as non-vapers.[4]

But there's one possible confounder that most of the studies did not look at. Teenagers are naturally more given to risk-taking and seeking excitement than us olds; those of us who were at one stage teenagers ourselves remember doing frankly ludicrous things that we would not dream of doing now, in our sedate later years.[5] Smoking weed and vaping both come under 'risky behaviour'.

And, of course, not all teenagers are equal. Some are more risk-averse than others. The ones who vape are likely to be the ones who also smoke cigarettes, drink alcohol or take drugs. This surely is not surprising to anyone.

Interestingly, two studies examined in the paper did look at something like this: they controlled for 'sensation-seeking', a personality trait defined by 'desire for stimulating, exciting, and novel kinds of experiences'.[6] People who score highly on the sensation-seeking scale, determined by a questionnaire, tend to be more interested in risky sports, driving fast and indulging in drink and recreational drugs. (Unsurprisingly, sensation-seeking peaks in the teens and early twenties and is higher in men than in women.)

The two studies that took sensation-seeking into account when looking at the link between vaping and marijuana use found different results from the others. One of them[7] found they were 1.9 times as likely, which is a lot lower than what

most of the others found; and the other[8] found no correlation at all (in fact, a slight decline). The fact that they tried to
control for sensation-seeking is probably part of the reason
why they found such a small result compared to some of the
others.

By controlling for potential confounders, it is possible to get
closer to the 'true' effect size. It's hard to be sure that if you've
controlled for the right things, or if you've missed something
out, or – as we'll discuss in Chapter 21, on collider bias – if you've
controlled for something you shouldn't have. It's complicated
and difficult.

None of this is to say that there's definitely no link between
vaping and cannabis use. You can tell some plausible stories: the
authors suggest that nicotine affects the developing brain and
actually makes it more sensation-seeking. Perhaps that's correct,
although it seems an implausibly large effect and the pre-existing
differences in sensation-seeking seem relevant.

But there's a general rule: when you see a news story saying
that X is linked to Y, don't necessarily assume that means that X
causes Y, or even vice versa. There could be some hidden thing,
Z, that causes both.

---

*You do not need to read or understand this box, but if you would
like to know how statistical regression works, go ahead.*

You may have heard the phrase 'statistical regression' before. It
sounds technical, but it's a fairly simple idea.

Let's imagine that you're interested in finding out whether how
tall people are is associated with how much they weigh. So we take
a large, random sample of people in the population, measure them
and weigh them, and put them all on a graph, one dot per person,
with their height on the X-axis and their weight on the Y-axis – that
is, taller people's dots are further to the right; heavier people's dots
are higher up. If someone is very short and light, they'll be at the

bottom left; if they're very tall and heavy, they'll be at the top right, and so on.

You look at the graph to see if there's a clear pattern. In this case, you can see that the data is sloping upwards – if someone is taller, they're probably heavier. This is what's called a 'positive association' (or 'positive correlation') – which just means that if one goes up, the other tends to go up too. If when one went up the other went down, we'd call it a 'negative association'. If the dots were all over the place with no clear trend, we'd say there was no association.

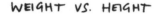

WEIGHT VS. HEIGHT

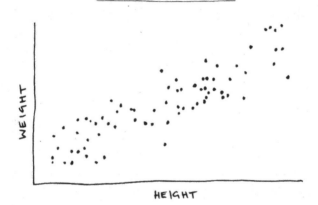

HEIGHT

Now, let's say we want to put a line through the data to describe the trend. How would we do that? You could draw it by eye, and you'd probably do a pretty good job. But there is a mathematically more precise way of doing it called the 'least squares method'.

Imagine you draw a straight line on the graph. It'll touch some dots, but most of them will be above or below the line. The vertical distance of each dot from the line is called the 'error', or the 'residual'. Take the value for each residual, square it (that is, multiply it by itself – that removes the problem that some values will be negative, because a number multiplied by itself is always positive), then add them all together. That figure is the 'sum of squared residuals'.

Whatever line you can draw that has the lowest possible sum of squared residuals is known as the 'line of best fit'. For the graph above, it would look like this:

WEIGHT VS. HEIGHT

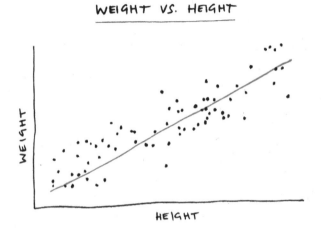

That line lets us make predictions – and the less residual there is (the lower the sum of the squares), the better those predictions will be. If we measure and weigh someone new, we would expect them to fall on or close to that line. Or if we know someone's height, we can predict their weight – for instance, let's say a height of 5'9" gives us a predicted weight of 12 stone, just by looking at the line. (You could also do vice versa: if you know someone's weight, you can guess their height, but you'd have to draw the line differently, measuring the horizontal error, and we probably don't need to get into it here.)

It should be noted that this probably won't give an especially accurate prediction of someone's weight, if you only use height. Other things – how much exercise you do, how much you drink, how many pies you eat in a week – will also help to predict weight. If you add those variables, then you can get a better picture of the true effect of height on weight. That's 'controlling' for other variables, as we discuss in this chapter. If you don't control for confounders, then you can end up over- or understating the correlation or finding apparent correlations where none in fact exists.

Chapter 8

Causality

Does drinking Coke make you get into fist-fights? Do you feel an uncontrollable urge to glass someone after downing an ice-cold Fanta?

Apparently, some people do, according to headlines in 2011. Those dreadful *young* people. 'Fizzy drinks make teenagers violent,' said the *Daily Telegraph*;[1] 'Fizzy drinks make teenagers more violent, say researchers,' echoed *The Times*.[2]

The headlines were based on a study in the journal *Injury Prevention*,[3] which found that 'adolescents who drank more than five cans of soft drinks per week . . . were significantly more likely to have carried a weapon and to have been violent with peers, family members and dates' – around 10 per cent more likely, in fact.

But it's worth noting the language here. The *Injury Prevention* study talks about Coke-drinkers *being more likely* to be violent. The newspapers said that the fizzy drinks *make* teenagers more violent.

There's an important difference there. The study finds a correlation, the sort of thing we've been talking about in the last couple of chapters: if one variable goes up, so does another. But as we've seen, that doesn't necessarily mean that the one going up *causes* the other to go up, any more than atmospheric carbon dioxide makes you fat, or ice cream sales cause drownings.

The newspapers use causal language. The fizzy drinks 'make teenagers violent' – implying they cause it; and, by extension, that if you take away the fizzy drinks, you will stop the violence.

As we've seen, it's very hard to tell even whether a correlation is direct – whether, when you take into account other things, ice cream sales really are correlated with drownings, or if they're both correlated with some other factor, such as air temperature. But often that's not really what we want to know. We want to know whether one thing *causes* another. How can we go about doing that?

Most of the studies we've looked at so far have been observational: that is, they look at the world as it exists. To go back to the carbon dioxide and obesity example, you look at how the levels of carbon dioxide in the atmosphere have changed, and you look at how deaths from obesity have changed, and you notice that they have both gone up.

The trouble is that this doesn't – it can't, really – tell you that CO_2 *causes* obesity (or deaths from obesity). It might be that increased obesity levels cause carbon dioxide to go up. Or (in this case more likely) it might be some confounding factor: perhaps that as countries get richer, they tend to both get fatter and emit more carbon, as we said in the last chapter.

There are ways of getting at what caused what with observational studies like these. For instance, causes usually have to come before effects: if you see that CO_2 levels go up before obesity does, that probably rules out an 'obesity causes carbon emissions' hypothesis. Another thing to look for is a 'dose response' – that is, the more of the hypothesised cause you see, the more of the effect you see. And, of course, it helps to have some good theoretical reason to believe that one causes the other. Wet pavements correlate with rainclouds; it is very easy to explain how the causal arrow might point one way, but rather harder to work out how the opposite might be true.

But unless it's staggeringly, unavoidably obvious – as in the case of rain causing wet pavements; or, more relevantly, in the case of smoking causing lung cancer, where the cause comes before the effect, there is a clear dose response, and a clear theoretical explanation, and the effect was so huge you couldn't ignore it – it will always be hard to establish causality with observational studies. So how do we go about working out whether one thing caused another thing?

Ideally, we use something called a *randomised controlled trial*, or RCT.

Here's the basic idea of RCTs. Let's use our earlier example of seeing whether this book makes people better at statistics. Except this time, instead of just looking at people who happened to have read the book, we're going to give it to them deliberately. You take a group of people – perhaps 1,000 of them. You make them all do a statistics test. Then you divide them up at random into two groups. One group is given the book to read; the other is given a placebo book, which looks just like this one except all the statistics are wrong. (If you do happen to find any errors in this book, you may have the placebo version.)

After both groups have read their books, you do the statistical test again, and you see whether either group or both groups have improved their average score. If *How to Read Numbers* makes people better at stats, then you'd expect to see that the group which has read it will have done better on average.

The control group is there to provide a 'counterfactual' – a sort of glimpse into an alternate universe. If you just test people before and after reading the book, and they've improved, it might be that the book made them better, but it might equally be that they've all been doing online courses at the same time. Or it might be that reading *any* book is enough to improve statistical ability. Or that just doing a study on people makes them behave differently.* So you create a control group to try to see what

---

* That's a problem in research, known as the 'Hawthorne effect', although there's some controversy over whether it's real. Between 1924 and 1927, a study was carried out on workers at the Hawthorne Works factory in Illinois. The study attempted to determine whether increasing the lighting at the plant made workers more productive. In popular retellings, the study found that literally any change – whether brighter or dimmer – improved output. But the original data was lost for a long time, and when it was found and reanalysed, no such effect was found. Others claim to have found it in other studies, but it remains controversial.

would have happened to the people you tested if they hadn't read the book.

Of course, you can't always do an RCT. Sometimes it's impractical, or unethical – you can't study the effects of smoking on children by giving 500 of them a pack of Embassy No.1s a day for ten years and comparing it to a control group, because that would be appalling. And you can't start wars in randomly selected countries to see the effect on their economies. Instead you can try to look for 'natural' experiments – places where groups have been separated at random for other purposes.

For instance, one famous study wanted to look at the effect on lifetime earnings of joining the military; but people who join the military are different from people who don't, so they couldn't just compare them.[4] Luckily (for the researchers, at least), in 1970, during the Vietnam War, the US military began to conscript soldiers. They were selected by a lottery – literally drawing balls from a machine like a bingo caller, live on television. It created what amounted to a treatment group (men who were drafted) and a control group (men who weren't). The study found that men who were drafted earned, on average, 15 per cent less over their lifetime than men who weren't.

Most observational studies, though, are not RCTs, and are not these randomised or quasi-randomised natural experiments. Most observational studies can simply tell you whether two or more numbers tend to go up and down at around the same time. They can tell you about correlation, but not about causation – and, as every pedant on social media will tell you, those aren't the same thing.

This is not always clear in newspaper reporting, however. One relevant paper looked at how seventy-seven observational studies (that is, non-RCTs, things that can't show you causality) were reported in the press, and found that almost half of them were reported with causal claims – that is, headlines like 'Daytime naps help improve learning in pre-school children', when the study only showed a link, not a causal link.[5]

*

Let's go back to the fizzy drinks. It won't surprise you now to learn that the study was observational – it did not give Irn-Bru to 500 teenagers, compare them with 500 teenagers who were given Diet Ribena, and then see which group was more likely to stab people at bus stops. Instead, it just looked at whether there was a link between how much fizzy pop they drank and how much violence they committed.

So we don't know whether the drinks caused the violence, or the violence caused the drinks (admittedly that sounds a little unlikely, although perhaps a street battle makes you work up a thirst) or – as in Chapter 7 – some other variable is linked to both. The study says that it controlled for various things, but the authors themselves suggest that there 'may be a direct cause-and-effect relationship', but that equally 'there may be other factors, unaccounted for in our analyses, that cause both high soft-drink consumption and aggression.' They did control for various factors – gender, age, alcohol consumption and others – but nonetheless, it can't show causality. It's just not reasonable to conclude, as the headlines did, that fizzy drinks cause violence, given that the study itself didn't make such a claim.

We're not saying that all RCTs are perfect – there are lots of practical issues that can mess them up, and they have a whole raft of problems of their own. But they are the most effective way of showing causality.

For readers, there's a simple rule of thumb here: if a study reported in the news isn't an RCT, then you ought to be very wary about any claims of causality. There might be excellent reasons to assume that the relationship is causal, but unless there's been some sort of randomisation, the study probably won't be able to tell you that.

*You do not need to read or understand this box, but if you would
like to know more about causality, go ahead.*

Sometimes researchers use a clever trick to establish causality with
observational studies, called the 'instrumental variable approach'.
Say you're an economist trying to study the impact of economic
growth on war in Africa. Obviously, conflict can reduce economic
growth, by stopping trade and investment and business. But that's
not the whole story. It might well be that reduced economic growth
can lead to a greater likelihood of conflict: with many people out of
work and angry, it's very easy to believe that a country might be at
greater risk of violence.

So if you observe that wars and economic collapse seem to go
hand in hand, how can you tell which is causing which?

If you think that A is causing B, but it in fact turns out that B is (or
is also) causing A, that's called 'reverse causality'. Of course, it can
be more complicated: A can be causing B, which is in turn causing A,
in a feedback loop. In the case of violence and economic growth, it's
easy to imagine how that might be the case – and if it is, then it will
interfere with your measurement just like a confounding variable.

So how can you tell the direction that the causal arrow is pointing? A → B, or B → A, or a loop? One way is to use an instrumental
variable – something you can measure which is correlated to one of
the things you're interested in, but not the other. In the case of wars
and economic growth, one such instrumental variable is rainfall.

A 2004 study tried looking at whether a slowing economy led
to war.[6] It found that a 5 per cent shrinking of the economy led to
a 12 per cent increased chance of war the following year. But, its
authors noted, even though the war followed the economic decline,
that doesn't prove a causal relationship. It could be that citizens,
aware that tensions are growing, change their behaviour, and shrink
the economy.

They decided, therefore, to look at rainfall. That might sound
strange, but rainfall is strongly linked to economic growth in

agriculture-based economies, where drought can lead to disaster; the higher the average rainfall, the more the economic growth. The assumption is that it's not strongly linked to war, except through the influence of the economy. So, if there are fewer wars in years with increased rainfall, that suggests that the economic situation really is influencing the likelihood of conflict, because rain will only affect war through the economy.

Lo and behold: the study found that the number of wars is lower in years with more rainfall, suggesting that the economy does affect conflict.

Of course, as with everything, it's *a bit more complicated than that*. You try to pick an instrumental variable which affects one thing but not the other, but it's hard to be sure that you've done it. In this instance, another economist points out that it's harder to wage war when your roads are flooded by heavy rains.[7] The researchers tried to take account of that, but whether they managed is not clear. This stuff is complex. Lots of academics get it wrong and it messes up their results, even if they're just looking for correlations.[8]

Chapter 9

Is That a Big Number?

You might remember a number that got written on the side of a bus at one point in the first half of 2016. It was quite a big number: £350 million. Apparently, we gave that sum to the EU every week; 'Let's fund our NHS instead,' exhorted the bus.

Don't worry, we're not going to relitigate the whole was-that-the-real-number thing. Various fact-checking outlets[1] and the UK Statistics Authority[2] agree that the real figure was more like £250 million – about £100 million never actually left our bank account, because of the rebate – and that, economically, we gained much more than that through trade, but it's not hugely important here. Instead, we're going to discuss whether it's a *big* number.

When is a number a big number? There's no such thing, really. Or rather: the bigness or otherwise of a number depends entirely on its context. A hundred is a big number of people to have in your house, but it's a small number of stars in the galaxy. Two is a small number of hairs on your head, but it's a huge number of lifetime Nobel Prizes, or gunshot wounds to the stomach.

Often, though, numbers in the news are presented without the context that you need to work out whether it's a big number or not. The most important piece of context is the *denominator*.

The denominator is the number below the line in a fraction: the 4 in ¾, the 8 in ⅝. (The number above the line is the 'numerator'.) It may not be a term that you've had much need for since maths classes at school, but it is vital when it comes to understanding numbers in the news. A large part of working out whether a number is big or not is to work out the best denominator for it.

Let's look at an example. In the years 1993 to 2017, 361 cyclists were killed on the road in London.[3] Is that a big number? It *sounds* quite big. But what's the denominator? A total of 361 cycle journeys ended in disaster in that twenty-five-year period. But how many were actually made in total? If we know what the bottom half of the fraction is, we can better understand the actual risk on any given ride.

Perhaps, since you're rarely given that information, it's assumed that you'll know it. If we asked you to guess, roughly how many cycle journeys do you think were taken on the average day in London between 1993 and 2017?

Imagine we told you that there were 4,000 of them. That would mean that there were about 36.5 million in the time period in question, which would equate to about one death in every 100,000 journeys.

Imagine we told you that there were 40,000. That would mean about one death in every million journeys.

Imagine we told you that there were actually 400,000 cycle journeys every day. Then it would be one death in every 10 million.

Which of these is right? If you don't know, you simply don't know the risk to cyclists who don their helmet to face the London streets. You don't know how big the number is: it's just orphaned, devoid of context. That's why providing the denominator is so important.

Let's put you out of your misery: the real figure, according to Transport for London, is about 437,000 daily journeys over that period. Whether a one-in-10-million risk of death per journey is too high is something only individuals can answer – but you can't answer it at all if you don't have the denominator.

(As an aside, it's worth noting that the average number of journeys has gone up hugely over the period covered – from 270,000 a day in 1993 to 721,000 in 2017. And in that time the number of *deaths* has decreased, bumpily but perceptibly: there were eighteen in 1993 and there were ten in 2017. So if you are a London cyclist, your risk of death on a given journey has fallen to roughly a sixth what it was in the early 1990s. And cycling is extremely good for you: even accounting for the risk of accidents and air pollution, you can expect it, on average, to increase your life expectancy significantly.[4])

The absence of a denominator is a common problem in news reports. In 2020 the *Daily Express* reported that 163 people had died in police custody over the previous ten years – but how many people had actually been in police custody?[5] If it was 1,000, that's a very different story to if it was 1 million. (According to Home Office statistics, it's closer to the latter: there are roughly a million *arrests* a year, although not all of them will end up in police custody.[6])

Crime is another example: if someone tells you that 300 people are murdered by undocumented immigrants in the USA every year, as Donald Trump said in 2018, you might think that sounds like a big number.[7] But is it really? What's the denominator?

In this case it's a little more complex – you need more than one. You could find out the number of murders in the US population as a whole: 17,250 in 2016, according to the FBI.[8] But that still doesn't tell us if it's a lot. We also need to know how many undocumented immigrants there are; then we can work out whether they are more or less likely to kill someone than the average US citizen.

Luckily, in 2018 the Cato Institute did some work looking into it. They found that in Texas in 2015 (which has a large illegal immigrant population), there were 22,797,819 'native-born Americans' and 1,758,199 'illegal immigrants', along with 2,913,096 'legal immigrants'.

They also found that native-born Americans committed 709 homicides, and that illegal immigrants committed forty-six. Those numbers allow us to divide the number of murders committed by each group by the number of people in that group – to divide our numerators by our denominators – and see which is bigger. In this case, 709 divided by 22,797,819 is 0.000031, or 3.1 murders per 100,000 people; 46 divided by 1,758,199 is 0.000026, or 2.6 murders per 100,000 people. So, in Texas at least, undocumented immigrants are less likely to be murderers than the average citizen. 'Legal', or 'documented', migrants committed about one murder per 100,000 people, in case you're interested.

Let's go back to the bus. That figure, £350 million, sounds huge. In many ways it is huge – it's hundreds of times more than the average person will earn in their lifetime. It might even buy you a four-bedroom house in north London.

But is it big? What's the denominator?

Let's take a look. First, £350 million multiplied by fifty-two is £18.2 billion. So that's how much we gave (according to the bus, at least; let's just stick with it) to the EU each year.

Total UK government spending, on everything from defence to road maintenance to pensions, was expected to be about £928 billion in the financial year 2020–2021, according to the 2020 Budget.[9] Dividing 18.2 by 928 (and then multiplying by 100, to get the percentage) gives us a little under 2 per cent. So £18.2 billion extra to spend would be about a 2 per cent increase to the national budget, for that year at least. (If you're still annoyed about it, if we'd used the £250 million figure it would be around a 1.4 per cent increase.)

That's not negligible: a 2 per cent increase to the national budget is equivalent to, for example, about half of our total spend on 'personal social services', i.e. local government support for vulnerable people, such as the elderly, the disabled, or at-risk children. But it might not be as staggering as it sounds. The trouble is, if you don't include denominators, you're just relying on people hearing a number and thinking it sounds big.

It might be a bit much to ask that all news stories quoting numbers try to find a suitable denominator for them. But as readers, it's worth asking yourselves, when you read some startling or impressive-sounding statistic: is this a big number?

Chapter 10

Bayes' Theorem

In the spring of 2020, when so many were locked down in their houses around the world, most of us were desperate for some sort of idea of how and when we could get out, and get society moving again. One plan that was mooted in several places, and widely[1] reported,[2] was the idea of 'immunity passports'.

The theory – which, at the time of writing, is still plausible, if unconfirmed – was that once you'd had the disease, you'd be immune to it: your body would develop antibodies that would fight off the disease, and they would protect you, if not for the rest of your life, at least for a prolonged period. The idea of immunity passports was that you'd be tested for those antibodies, and then – if you tested positive – you'd be given a certificate saying that you had had the disease, were immune and could restart your life. You would not be at risk of either catching the disease or spreading it to others.

Of course, whether the passports would work depends on how accurate the tests are. But at the time of the stories we mention, the US FDA had already issued emergency approval for a test which claimed that it would be right 95 per cent of the time.[3] So if you took a test and got a positive result, what's the likelihood that you were immune? It's about 95 per cent, right?

Nope. If that's all the information you have, then the answer is *you have absolutely no idea*. There is not enough information given to provide you with the faintest clue what your chances of being immune are.

This is to do with something called Bayes' theorem, named for the Reverend Thomas Bayes, an eighteenth-century Presbyterian clergyman and mathematical enthusiast. It's a simple piece of reasoning, but it throws up some extremely strange results.

Bayes' theorem, when written in logical notation, looks kind of scary: $P(A|B)=(P(B|A)P(A))/P(B)$. But it's actually pretty straightforward. What it describes is the probability of a given statement, A, being true in the event that another statement, B, is true – you can see the box below for more details. What makes it important and counterintuitive is that it takes into account the *prior probability* of statement A being true, before you know whether B is true or not.

You do not need to read or understand this box, but if you would like to know more about conditional probability, go ahead.

Bayes' theorem is about conditional probability, which you may remember from school. Imagine you have a freshly shuffled pack of cards. What are the odds that the first card you draw is an ace? It's 4/52, because there are four aces in the fifty-two-card deck. And because both 4 and 52 divide by 4, you can write it as 1/13.

Let's say you draw an ace on the first go. What are your chances of drawing an ace on your second card? You've already got one of the aces, and one of the cards has gone, so your numbers have changed: it's now three aces in fifty-one, or 3/51.

That is your probability of drawing an ace, *conditional* on the fact that you've already drawn an ace and discarded it.

In statistics, the probability (let's call it P) of an event (let's call it A) is written like this:

$$P(A)$$

If we have another event which happens before A – let's call it B – then we write it like this:

$$P(A|B)$$

That vertical line symbol | means 'conditional on'. $P(A|B)$ simply means 'the probability of A, conditional on the fact that B has already happened'. So P(A|B) for 'drawing an ace, conditional on the fact that you have already drawn an ace and discarded it' is 3/51, or about 0.06.

This is quite hard to explain with the notation alone, but it's easier if we look at an example. The most famous examples are usually in medical screening. Imagine you have a blood test which can detect a rare but deadly neurodegenerative disease early in its course. It's immensely accurate.

Importantly, there are two kinds of accuracy: how likely it is to correctly tell someone who has the disease that they have the disease, its true positive rate or 'sensitivity'; and how likely it is to correctly tell someone who doesn't have the disease that they don't have the disease, its true negative rate or 'specificity'. We're going to imagine that it scores 99 per cent on both.

But – and this is vital – the disease is very rare. Let's say one in 10,000 people has it at any one time. That's your prior probability.

So you give a million people your test. Given that one in 10,000 have it, that means there will be 100 actual cases of the disease. Your test will correctly tell ninety-nine of them that they have the disease. So far, so good. It will also correctly tell 989,901 people that they don't have the disease. Still looking pretty good at this stage.

There is, though, a snag. Even though it's got it right 99 per cent of the time, it's still told 9,999 perfectly healthy people that they have a fatal illness. Of the 10,098 people it's told have the disease, just ninety-nine of them actually do; less than 1 per cent. If you were to take this test at face value, and told everyone who got a positive result that they were ill, then you would be getting it wrong (and scaring people, and perhaps getting them sent for unnecessary, intrusive and risky medical procedures) ninety-nine times out of every 100.

Without knowing the prior probability, you can't possibly know the meaning of a positive test. It can't tell you how likely you are to have the disease it's testing for. So reporting a number like '95 per cent accurate' is meaningless.

This is not a hypothetical problem of interest only to academics. One meta-analysis (a paper which aggregates the results of other studies, you'll remember from Chapter 7) found that 60 per cent of women who have annual mammograms for ten years have had at least one false positive.[4] A study which looked at men who had been sent for biopsies and rectal examination after a positive prostate cancer test found that 70 per cent of them were false positives.[5] One antenatal screening test for foetal chromosomal disorders which claimed 'detection rates of up to 99 per cent and false positive rates as low as 0.1 per cent' would actually, given the rarity of the conditions involved, have returned false positives between 45 per cent and 94 per cent of the time, according to one paper.[6]

It's not that these tests are taken as definitive – people with a positive result will be given more comprehensive diagnostic examinations – but they will scare a lot of patients who don't, in the end, turn out to have cancer, or foetal abnormalities.

And it's not just medical testing. It has vital implications in law as well. In fact, a well-known and common failure of the law courts, the 'prosecutor's fallacy', is essentially a misunderstanding of Bayes' theorem.

Andrew Deen was convicted of rape in 1990, partly on the basis of DNA evidence, and sentenced to sixteen years in prison. A forensic expert appearing for the prosecution said that the probability that the DNA came from someone else was one in 3 million.[7]

But as Chief Justice Lord Taylor pointed out in a review of the case,[8] this was mixing up two distinct questions: first, how likely would it be that a person matched the DNA profile if they were innocent; and second, how likely would they be to be innocent if they matched the DNA profile? The 'prosecutor's fallacy' is treating those two questions as if they are the same.

We can do the exact same thing as we did with the medical

test. In the unlikely event that you have no other evidence – you have simply picked your suspect from the entire British population, at the time around 60 million – then your prior probability that any random person is the murderer is one in 60 million. If you run your test on all 60 million people, it will correctly identify the murderer, but it will also return false positives on twenty innocent people. So even if there's only a one in 3 million chance that you'd see a positive result if an innocent person took the test, there is a greater than 95 per cent chance that a randomly selected person who got a positive result would be innocent.

In reality, defendants are not picked at random; there is usually other evidence to support a case, meaning that the prior probability is greater than one in 60 million. But, just as with medical tests, knowing the probability of a false positive on DNA evidence doesn't tell you how likely it is that someone is innocent: you have to have a prior probability, some assessment of the chance that they were guilty to begin with.

In December 1993, the Court of Appeal quashed Deen's conviction, declaring it unsafe because, they said, the judge and the forensic scientist had both fallen prey to the prosecutor's fallacy. (He was, however, once again convicted in the ensuing retrial.)

Similarly, the tragic case of Sally Clarke – convicted in 1998 of murdering her children, because an expert witness said that the chance of two babies dying of Sudden Infant Death Syndrome in one family was one in 73 million – turned on the prosecutor's fallacy: the witness failed to take into account the prior probability of someone being a double murderer, which is even rarer than SIDS.[9] (There were other problems, notably that the expert witness didn't account for the fact that families that have already had one case of SIDS are more likely to have another.) Clark's case was also overturned, in 2003.

Where does this leave us with the immunity passports? Well – if you get a positive result on your antibody test, even if it is 95 per cent sensitive and 95 per cent specific, you don't actually know how likely you are to have had the disease. It is all about how likely it is, before you take the test, that you might have the disease – your prior probability. The most obvious starting point for that is the prevalence of the disease in the population.

If 60 per cent of the population have had it, then if you test a million people, 600,000 of them will have had it and 400,000 won't; your test will correctly identify 570,000 of the people who have, and incorrectly tell you that 20,000 people have had it when they really haven't. So your chance of having a false positive, if you've had a positive antibody test, would be just 3 per cent.

But if only 10 per cent of the population have had it, then of your million people 100,000 people will have had it, of whom your test will correctly identify 95,000; but of the remaining 900,000, it will tell 45,000 of them that they had. If you tested positive, there would be a 32 per cent chance that you hadn't had the disease – but you would think you were now safe to go out on the streets, or to visit elderly grandparents, or work in a care home.

Again, these sums only apply if you are testing the population at random. You could get a better estimate if, say, you only gave the test to people who had had the core symptoms. Then you'd be testing a population that was more likely to have the disease, so a positive test would be much better evidence. Your prior probability would be higher. But unless you have some estimate of the prior probability, you can't know what your test means.

This is a difficult concept to get your head around – and not only for readers, or journalists. A 2013 study asked almost 5,000 US obstetrics and gynaecology residents – that is, qualified doctors – to work out the likelihood that someone would have cancer, if 1 per cent of the population had the disease and they had received a positive result in a 90 per cent accurate test.[10] The correct answer was about 10 per cent, but even on a multiple-choice questionnaire 74 per cent of doctors got it wrong.

Still, it's important. It's important because we see stories about screening, and tests for disease, and so on – and without this information, it might seem as if a positive result in a 95 per cent accurate test means you're 95 per cent likely to have the disease. But it doesn't. If you see stories about '99 per cent accurate tests', whether they're about cancer screening, DNA profiling, Covid-19 tests or anything else, treat them with caution if they don't address these issues.

Chapter 11

Absolute vs Relative Risk

Scary news for older fathers in the *Daily Telegraph* in 2018: men who have children at forty-five or older 'are more likely to have babies born with health problems'.[1] Specifically, children born to older men 'had 18 per cent higher odds of having seizures, compared with youngsters with fathers aged 25 to 34 years', among some other things. It made, to be fair, a welcome change from the usual scaremongering about older mothers, and the (usually hugely overstated) higher risks of infertility and various birth defects.

The story was based on a study published in the *BMJ*, which looked at the impact of paternal age on infants' outcomes.[2] And it did indeed find the sort of increased risk mentioned.

But there's something that the *Telegraph* story *didn't* mention – 18 per cent more than what?

When you see that something has increased by 75 per cent, or gone down by 32 per cent, or whatever, that is a *relative* change. If we're talking about risks – if you eat five or more roasted swans a week, your lifetime chance of getting gout goes up by 44 per cent, that sort of thing – then we are talking about *relative risks*.

You'll see risks presented like that a lot. In 2019, for instance, CNN reported that bacon increases your risk of bowel cancer: the more you eat, the higher your risk, with the risk 'rising 20% with every 25 grams of processed meat (roughly equivalent to a thin slice of bacon) people ate per day'.[3]

Or, to return to the risk of paternal age and birth defects, in

2015 it was reported that teenage fathers were more likely – 30 per cent more likely, according to the *Daily Mail* – to have children with 'autism, schizophrenia and spina bifida'.[4]

An increase of 30 per cent sounds scary. Or an increase of 20 per cent, or 18 per cent. These are all significant-sounding numbers. It may even sound as though your risk of getting bowel cancer *will be* 20 per cent, or your risk of your child having spina bifida *will be* 30 per cent if you eat bacon or have a baby while under the age of twenty.

Of course that isn't what it means. A 30 per cent increase in risk means that your risk goes up from some level, X, to 1.3 times X. But unless you know what X is, that doesn't help you a great deal. That's why it's important to present these things in terms of *absolute risk*. That is: how likely it actually is that something will happen, not simply how much it has changed.

In the case of the risk of bowel cancer for bacon-eaters, a British person's background risk of developing bowel cancer in their lifetime is about 7 per cent for men and about 6 per cent for women, according to the charity Cancer Research UK.[5]

Obviously, that is not negligible – somewhere around a one in fifteen chance of developing the disease, depending on your sex. But now look at what a 20 per cent increase means.

Let's take the largest estimate. Say you're a man in the UK. Your background risk of bowel cancer is about 7 per cent. You eat an extra rasher of bacon every day (about 25g). That puts your risk up by 20 per cent.

But remember – that's 20 per cent *of 7 per cent*, which is 1.4 per cent. So it goes up from 7 per cent to 8.4 per cent. If you're unwary, or unused to dealing with percentages, you might think it goes up by 20 percentage *points*, or up to 27 per cent. But it doesn't.

So your risk of getting bowel cancer goes up from about one in fifteen to about one in twelve. It's not nothing, but it sounds much less scary than 'a 20% increased risk'.

You can take it even further than that, in fact. We would expect about seven out of every 100 British men to get bowel cancer at some point in their lifetimes. If they all start eating an extra rasher of bacon a day, about 8.4 of them will get bowel cancer, instead of seven. That extra ration of bacon a day translates to about a one-in-seventy chance that you will develop a cancer that you otherwise would not have. If you're a woman, then the chances are even smaller.

That's not to say that a one-in-seventy chance of getting cancer is negligible. It's important information which you can use to help you decide whether or not to change your diet. But it's a totally different thing to 'a 20% increase', which can tell you nothing at all about the risk you run. It's a trade-off between the benefits you gain from eating extra bacon – it's nice to eat bacon! It might make your life more enjoyable! – and the risk. In order to work out whether the trade-off is worth it, you need good information.

Relative risk can also be used to make drugs seem more effective than they are – for instance, one cancer drug in the USA declared in its adverts that it 'reduced the risk of dying by 41% compared to chemotherapy', which sounds good: but it actually translated into an average of 3.2 extra months of life.[6] Research by the US FDA found that when doctors are given the 'relative effect measures' of drugs, rather than the absolute effects, it was 'associated with greater perceptions of medication effectiveness and intent to prescribe' – that is, even doctors are fooled by relative risk.[7] Presenting numbers in absolute terms allows all of us, patients and doctors, to better understand the dangers.

Relatedly, be wary when you read that something – a political party, perhaps, or a religion – is 'fast-growing'. It may be that a political party is indeed the fastest-growing in relative terms, if it has doubled in size in a week; but if you were to learn that last week it had one member, and this week that one member recruited her husband to join, so now it has doubled in size to two members, you might not be so impressed.

*

Let's go back to the original story about older fathers and children who have seizures. You know the relative risk increase – 18 per cent. But by now you know that that, on its own, doesn't tell us very much. What we need to know, instead, is the *absolute risk*: how likely your child actually is to have seizures if you become a father in later life, compared to how likely they would have been if you had had the child earlier.

The numbers you need are 0.024 per cent and 0.028 per cent. The risk of your child having seizures, if you become a father between the ages of twenty-five and thirty-four, is twenty-four in 100,000; the risk, if you become a father between the ages of forty-five and fifty-four, is twenty-eight in 100,000. The difference will affect, on average, four babies out of every 100,000 born.

None of this is to say that the difference is unimportant. Even a four-in-100,000 chance is a real chance. But it has to be weighed against the trade-offs. Some people will want to have a child later in life; they may consider it worth the small extra risk.

All that said, it is hard to blame the media entirely for this. Many scientific papers fail to give absolute risk, despite most journals saying in their guidelines that they ought to. The *BMJ* article about older fathers, for instance, reported all its findings in relative risk, against *BMJ* guidelines. And when the studies themselves do, sometimes the press releases do not; journalists, usually pushed for time and often not statistically literate, will struggle to find the information in the paper itself (if it's actually there, which it may well not be), or perhaps will not realise that there's a need for it, even if they have access to the study.

But it is a crucial aspect of communication. The role of science journalism, at least when it comes to things like lifestyle risks, surely has to be to provide useful information for readers – if I have a glass of wine a night, will that give me cancer, or heart disease? That information needs to be presented in absolute terms, or it is of no use. Scientific journals, university press offices and the media all need to establish, as an immutable rule, that risk should be presented in absolute, not just relative, terms.

Chapter 12

Has What We're Measuring Changed?

'Hate crimes double in five years in England and Wales,' the *Guardian* reported in October 2019.[1] Which sounds extremely bad.

The headline referred to statistics of hate crimes reported to police between 2013 and 2019.[2] It reported – correctly – that 103,379 hate crimes were recorded in 2018–19, 78,991 of them race-related. That's up from 42,255 in 2012–13.

You may or may not be surprised to read that, and either response would be reasonable. We live in an age of dreadful and high-profile hate crimes; but there is also an overall societal trend towards lower levels of prejudice. The British Social Attitudes survey, for instance, has found that acceptance of same-sex relationships has increased: less than 20 per cent of Britons said they think they are 'not wrong at all' in 1983, but that number had risen to more than 60 per cent by 2016.[3] Similarly, in 1983, more than half of white Britons said they would mind if a close relative married a black or Asian person; by 2013, that had dropped to 20 per cent.[4]

It's perfectly possible that social attitudes might improve on average while the prejudiced minority become more extreme. But it is nonetheless surprising that hate crimes appear to have doubled, when the number of people holding the attitudes presumably behind those hate crimes has dropped by well over half. What's going on?

Let's talk about something else first. Diagnoses of autism, a developmental disorder which involves problems with social communication and interaction, have been on the rise for years. In 2000, the US Centers for Disease Control and Prevention estimated that about one in every 150 children had autism-spectrum disorder; by 2016, that figure was one in fifty-four.[5] Even the 2000 figure was itself many times higher than estimates in previous decades: studies in the 1960s[6] and 1970s[7] suggested that only about one child in every 2,500 or even 5,000 was autistic. Similar trends can be seen in countries around the world, especially rich ones.

Figures like that have led to reports of an 'autism epidemic'. It is traced back to various causes; psychiatrists blamed cold and distant parents ('refrigerator mothers', in their awful phrase).[8] That turned out to be entirely false, and there could be all sorts of reasons why emotionally distant parents might tend to have emotionally distant children. Later, heavy metal contamination, herbicides, electromagnetic radiation, gluten, casein and, of course, vaccines were all floated as possible explanations.

But none of these make sense as explanations. We don't use more herbicides than we used to, and there's no evidence that glyphosate, the herbicide most commonly named as a suspect, has any link to developmental disorders. There's no plausible mechanism or epidemiological evidence for a link between radiation and autism. The vaccine theories have no evidence to support them – and besides, if a given vaccine caused autism, you'd expect to see spikes in national autism diagnoses shortly after that vaccine's introduction, and we simply don't. In fact, no one has been able to find any convincing environmental risk factors for autism; it seems to be mainly a combination of heredity and randomness.

So how come autism diagnoses rose so astonishingly quickly?

Here's what seems to have happened.[9] The second edition of the Diagnostic and Statistical Manual of Mental Disorders (DSM-II), published in 1952, included no diagnosis of 'autism' at all: the word was mentioned solely under childhood schizophrenia. In 1980, the third edition (DSM-III) was released, in which autism was given its own diagnosis; it was described as a 'pervasive developmental disorder', rooted in brain development. It gave criteria for the diagnosis of autism, including 'lack of responsiveness to other people', 'gross deficits in language developments' and 'bizarre responses to the environment'.[10] If a child met those criteria and was seen to do so before the age of thirty months, they were diagnosed with autism.

DSM-III was revised in 1987, broadening the diagnosis to include milder cases, giving a list of sixteen criteria on which to diagnose (children had to meet eight of them), and allowing children over thirty months to be diagnosed. For the first time autism was broken into two parts, 'autism' and 'pervasive developmental disorder – not otherwise specified' (PDD-NOS), meaning that children who didn't meet the full definition of autism but still required support could receive a diagnosis.

DSM-IV, published in 1994, was the first to use the word 'spectrum', and included five different conditions, including the well-known Asperger's.

The current edition, DSM-5 (they stopped using Roman numerals, for some reason), removed the separate diagnoses altogether, and lumped three of them together under 'autistic spectrum disorder', with no clear distinctions between them. (The other two were moved out of the autism category.)

So what it meant to be 'autistic' changed repeatedly over the course of several decades, from there being no separate condition called 'autism' at all to there being five, to there being one extremely widely defined one. In that time, it expanded; children who would not have been included in earlier definitions were diagnosed under later ones.

Suddenly we have a simple explanation for why autism diagnoses went up so much: the term 'autism' changed its meaning several times, expanding to include more people. Plus, as the condition became more widely known both by parents and doctors, and as meaningful ways of improving the lives of autistic children became available, more children were screened to see whether they met the criteria.

It may well be that nothing changed, in terms of the distribution and prevalence in the population of the psychological traits that we now associate with 'autism'; the growth in the apparent autism rate could have come from the fact that the medical establishment changed what it was measuring, and became more careful in looking out for the sort of thing that could be autism.

Changes in how statistics are recorded can hugely affect the apparent trend in those statistics. For instance, between 2002 and 2019 the number of sexual assaults recorded by police in England and Wales tripled, from around 50,000 to about 150,000.[11] But that's because, historically, the police and courts have been bad at taking sexual crimes seriously (startlingly, marital rape was not a crime until 1991).[12] Societal change has pressured the police to do better, and now they are more likely to record them.

If we wanted to compare 2002 and 2019 sexual assault rates using police data, then we'd need to have some sort of way of seeing how many the police *would* have recorded in 2019, had they used the same methods, attitudes and criteria as they did in 2002. That's not possible. But there are other things we can look at.

The Crime Survey for England and Wales (CSEW) is a big survey of the population which asks people how often they have been victims of crime. It's intended to determine trends in crime rates, so it's maintained a consistent methodology for decades. It is, therefore, not subject to changes in police recording habits, although, of course, it may be subject to changes in public behaviour, such as if people have become more comfortable talking about and reporting sexual crimes – which, historically, they have not been, for many reasons. It records subtly different things from the police-recorded data, but it should still reflect a similar underlying reality.

And the CSEW found that the number of sexual assaults that had actually taken place over that period had dropped, from about 800,000 in 2004 to about 700,000 in 2018.[13] The changes in how data was recorded and measured seems to have taken a trend that was going one way and made it look as though it was going the opposite way. (That said, it is important to note that the CSEW data only looked at assaults on people aged sixteen to fifty-nine, while the police-recorded data included those on children and older people; we don't think that will materially change the point, but it does mean it's not quite looking at the same thing.)

Measurement and recording practices change quite regularly, often for good reason. It happened repeatedly during the early months of the Covid-19 outbreak. For a long time, most US states only counted a death as Covid-related if there had been a lab-confirmed positive test. Then, on 26 June 2020, several states agreed to include 'probable' deaths, i.e. those in which the patient had the symptoms but not an actual test to confirm that they had Covid, because it was clear that the test-confirmed numbers were missing a large fraction of the real deaths. So on 26 June, there was an apparent spike in the death rate, even though, in underlying reality, nothing had actually changed.[14]

So what's gone on with the hate-crime data? Just as with sexual assaults, the headline figure was reporting police-recorded crime. And, just as with sexual assaults, the police may not, historically, have been especially brilliant at taking hate crimes – whether race-, gender-identity-, disability- or sexuality-related – as seriously as they should. In recent years, mercifully, that has begun to change.

Also, as with sexual assaults, we can't look at how the police *would* have recorded the data if they'd been using their 2013 methods and attitudes. But we can use the Crime Survey for England and Wales data again – which, you'll remember, is a giant survey of the population, letting us see the prevalence of different kinds of crime without relying on police-recording methods.

Again, these numbers aren't directly comparable, because the CSEW data doesn't include quite the same things as the police-recorded data; but it does show that the real trend appears to be in the opposite direction. The CSEW found that there were about 184,000 hate-crime incidents in 2017–18, and that that was down from about 300,000 in 2007 and about 220,000 in 2013.[15] The *Guardian*, it should be acknowledged, did mention that the apparent increase was 'partly because of improvements in crime recording'.

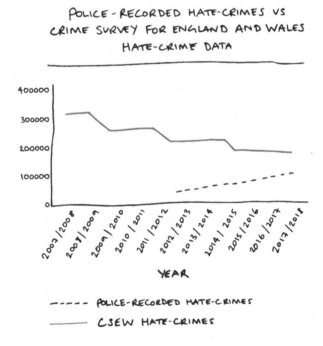

POLICE - RECORDED HATE-CRIMES VS
CRIME SURVEY FOR ENGLAND AND WALES
HATE-CRIME DATA

- - - - - POLICE-RECORDED HATE-CRIMES
————— CSEW HATE-CRIMES

None of this makes any of it all right: 184,000 is still an awful lot. And the CSEW did see some evidence of real spikes after the 2016 referendum and the series of terror attacks in 2017. But it does go to show that changes in how things are recorded and measured can entirely turn the story on its head, from one in which a number is going down to one in which it is going up. And if the media isn't careful to make that clear, you can get a backwards impression of what is going on.

Chapter 13

Rankings

The 'UK rises in international school rankings', said a headline on the BBC website in 2019.[1] According to the PISA rankings, which compare the educational performance of children around the world, the UK had gone up from twenty-second to fourteenth on reading skills in the space of a year; it had also gone up the rankings in science and in maths. Which sounds good, right?

Well, obviously it's probably not bad (for the UK, at least, although for one country to go up, another must go down). But crude rankings like this can conceal a lot of information. All they do is put a list of numbers in order from largest to smallest; they tell you who comes first, second and third (and last). But on its own that doesn't tell you a great deal, if you don't care about those rankings for their own sake.

For instance, we read a lot that the UK is the 'fifth-largest economy in the world'. Or at least we used to. In 2019, according to the IMF, it was leapfrogged by India.[2] This was a shame from the point of view of some British people, who had staked a surprising amount of national pride on this position in the rankings. (This is not the first time this has happened. Britain, France and India have swapped places on the IMF table several times over the last few years; France held the fifth spot in 2017,[3] India held it again in 2016.[4])

But what difference does it really make, from Britain's point of view, if they are fifth, sixth or seventh? What does the position in the rankings tell you about Britain's economy?

Obviously, you can tell that it has not been growing as fast as

India's, in the year since the rankings were last released. But does it mean that Britain's economy is large? You'd think that, since there are 195 countries in the world, the fifth-biggest would be quite big, but is that true?

To use a football analogy: in the 2018–19 season, Manchester City finished first and Liverpool finished second. In 2019–20, Liverpool (eventually, after a three-month Covid-19-enforced break) finished first and City finished second. If the rankings are all you have to go on, then you might think that those two seasons were pretty similar. But the rankings conceal a major difference: in 2018–19 City finished one point ahead of Liverpool; in 2019–20 Liverpool finished eighteen points ahead of City.

Similarly, according to the IMF rankings, the top seven countries by raw GDP are the USA, China, Japan, Germany, India, the UK and France.[5] Is it a photo finish, like in 2018–19, or an almighty thrashing, as in 2019–20?

Let's take a look.

RANKINGS

COUNTRIES	RANK 2019	GDP $US MILLIONS	SHARE OF WORLD GDP
UNITED STATES	1	21,439,453	24.57%
CHINA	2	14,140,163	16.20%
JAPAN	3	5,154,475	5.91%
GERMANY	4	3,863,344	4.43%
INDIA	5	2,935,570	3.36%
UNITED KINGDOM	6	2,743,586	3.14%
FRANCE	7	2,707,074	3.10%

The UK and France are almost indistinguishably close – the UK's economy is just 1.3 per cent bigger than France's – and measuring a country's economy is tricky so it's probably within the margin for error. India's is slightly bigger again – about 7 per cent bigger than the UK's – but it's hardly overwhelming.

But when you jump up a rung, you get Germany, which is 40 per cent bigger than the UK. Japan is 87 per cent bigger. And China and the USA aren't even on the same playing field. China's economy is 380 per cent bigger than the UK's – almost five times the size – and the USA's is 630 per cent larger, more than seven times the UK's. The argument over who gets to come fifth is very much Everton, Arsenal and Wolves scrapping for a Europa League spot.

We can also answer that question about whether the UK's economy is very big, in the scale of things. As a share of world GDP, the USA is genuinely enormous: almost one dollar in every four spent in the whole world goes through American hands. About one in six goes through Chinese ones. The UK, on the other hand, makes up a little over 3 per cent of the global economy. For comparison, Virgin Cola, the drink that Sir Richard Branson launched in the early 1990s in Pamela Anderson-shaped bottles in an attempt to rival Coke and Pepsi, successfully managed to corner about 3 per cent of the UK market in cola-flavoured soft drinks and was discontinued after a few years.[6] Virgin Cola may well have been the third-biggest cola drink in the UK, but it was still not very big; the UK may have been the fifth-biggest economy in the world, but that doesn't mean very much.

We're *still* missing out on a lot of information, though. Imagine that tomorrow someone invents something – let's say cold-fusion energy generation using two lemons and an empty Fanta can. Overnight, every economy in the world grows by ten times.

We look at our table, and Britain is still sixth, behind India. It just has an extra zero on the end of its GDP figure.

It's true that relative wealth is important, and there is evidence that we gain in happiness at least partly by how wealthy we are in comparison to others, not in an absolute sense.[7] But our Fanta-can cold-fusion invention would revolutionise the world; it would bring hundreds of millions out of poverty. And yet as far as our rankings are concerned, nothing has changed. France is still slumming it in seventh, the useless strike-happy layabouts.

(It's also worth noting that you don't really care, individually, about how big your country's GDP is when you add all the people together. Liechtenstein will always have a tiny GDP because there aren't many people there, but most of those people are quite well off; meanwhile, Indonesia has quite a big GDP, because it has a lot of people, but many of them are quite poor. GDP *per capita* is probably more interesting; the IMF has the UK significantly further down that list, in twenty-first place.[8])

It's not that rankings are entirely without value. They can tell you something about how you're doing relative to your peers, whether 'you' are a sales employee, a school in Leicestershire or a medium-sized western European democracy. It might, for instance, be useful to know whether Britain is lagging behind Germany on Covid-19 swab testing, or where our spending on the arts or on national defence lies in comparison to other countries. But even so, it's only useful if we are *also* told what data the ranking is based on. If we're lagging behind Germany on swab testing in that they're carrying out 500 per 100,000 population while we're carrying out 499, we probably don't care; if it's 500 to fifty, then something might have gone wrong somewhere.

But we love to quantify everything these days – university rankings, school rankings, hospital rankings. Curry-house rankings and kebab awards.

There's an added problem, which is that a lot of rankings are based on collected subjective opinions. The World University Rankings score, for instance, is heavily based on 'academic reputation', with 40 per cent of a university's score dependent on it.[9] Academics are surveyed and asked how good they think the teaching and research is at 200 different universities. Since most of those academics will never have attended a lecture at most of those universities, there will be a lot of guesswork going on; correspondingly, the rankings are quite volatile. The University of Manchester, for instance, where David studied, is twenty-seventh in the World University Rankings, but fortieth in the *Guardian*'s list of UK universities.[10] That is, obviously, ridiculous: if there are thirty-nine better universities in the UK, there can't be only twenty-six in the whole world, because the world contains the UK. King's College London, where Tom did a postgrad degree, is also weird: sixty-third best university in the UK, but thirty-first best in the world.

These counterintuitive outcomes are down to decisions about what to include and how much weight to give them: if you make 'student satisfaction' more important than 'academic reputation', you get different results. Arbitrary decisions about what to take into account change things enormously. That doesn't automatically make them *wrong*, but rankings shouldn't be seen as some sort of holy truth.

So, the PISA rankings. What are they based on? Are the rankings of much use?

First, let's acknowledge that they're not as subjective as university rankings. The scores are based on standardised exams given to fifteen-year-old children in all the countries ranked; the questions cover maths, science and reading skills. And these tests seem to have real-world validity: children who do well in the PISA test tend to progress further in education and are more likely to end up in employment in later life than the children who do less well.[11] That implies that the test is measuring something real, so the rankings aren't entirely meaningless.

But the PISA *rankings* are based on the PISA *scores*, and most rich developed democracies like ours have very similar PISA scores. For instance, looking at reading: the UK's average score is 504, the same as Japan's; one better than Australia and one worse than the USA.[12] The scores range all the way from 555 (in four

Chinese provinces) to 320 (Mexico and the Philippines); twenty countries, almost all of them rich, developed democracies, are crammed in between 493 and 524 points. Even a small, statistically insignificant change would cause the UK to drop several places. In fact, PISA helpfully tells us that the UK's score is statistically indistinguishable from those of Sweden (on 506 points), and New Zealand, the USA, Japan, Australia, Taiwan, Denmark, Norway and Germany (on 498 points). A country could, in theory, have jumped from twentieth to eleventh place without anything actually changing. (The UK's ranking in maths went up from twenty-seventh to eighteenth, and that *was* statistically significant, apparently.)

Again, this doesn't mean that rankings are useless. But it does mean that the rankings on their own are not very helpful: you need to know the scores that are used to make them, and how those scores are made. You care if your football team finishes one point above its rivals; you may well not care at all if your economy is 1 per cent smaller than India's.

Chapter 14

Is It Representative of the Literature?

Hey, good news! 'Drinking a small glass of red wine a day could help avoid age-related health problems like diabetes, Alzheimer's and heart disease, study finds.'[1]

But hang on! 'A glass of red is NOT good for the heart: Scientists debunk the myth that drinking in moderation has health benefits.'[2]

Hmm.

Hey, more good news! 'One glass of antioxidant-rich red wine a day slashes men's risk of prostate cancer by more than 10 per cent.'[3]

But hang on again! 'Even one glass of wine a day raises the risk of cancer: Alarming study reveals booze is linked to at least SEVEN forms of the disease.'[4]

Yes, it's a rollercoaster ride being a red-wine drinker and a reader of the *Daily Mail*. These headlines are all based on real studies from the last five years; it's not that the *Mail* is making anything up (or that the *Mail* is uniquely prone to doing this). So what's going on here? Does red wine make us live for ever, or is it killing us?

Think back to Chapter 3, when we talked about sample sizes, and Chapter 5, when we discussed p-values. If you're doing a study, or an opinion poll, or anything where you're trying to use sampling methods to try to find out something – how many people are likely to vote Labour, how effective a drug is in treating a disease – the answer you get will not necessarily be the exact truth. Even if you've got an unbiased sample, and the study is conducted well, the number you get back could randomly be higher, or lower, than the real truth, just through the workings of chance.

This has an obvious implication. Imagine that eating fish fingers slightly reduces the risk of snoring. (Admittedly a very unlikely scenario, but let's imagine it anyway.)

Let's say that there have been lots of different studies into whether or not fish fingers affect snoring. And let's say that, while some of the studies are quite small, they've all been perfectly well conducted, and there's no publication bias (see Chapter 15), p-hacking (Chapter 6), or other dodgy statistical practice going on. (This too is a very unlikely scenario. But let's keep going.)

What we'd expect is that the *average* finding of the studies would be that fish-finger-eaters snore slightly less. But *any individual study* could end up returning slightly different results. If the studies truly are unbiased, then you'd expect them to cluster around the true effect in a normal distribution (which you'll remember from Chapter 3). Some will be higher, some lower, most of them about right.

So if there have been lots of studies into the fish-finger/ snoring link, some of them will return results that aren't representative of reality. They might overstate or understate the effect; they might even find that there isn't one, or that fish fingers *cause* snoring. Again, there needn't be anything wrong with any of the studies, or the publication process. This is all just the workings of randomness.

The virtuous thing to do is to try to work out what all these studies are clustering around: what the average result is. This is why people do literature reviews at the beginning of academic papers – to put their results into the context of the scientific literature as a whole. Sometimes researchers do meta-analyses – academic papers which go through all the existing literature and try to synthesise the results. If there have been enough studies, and if there isn't a systematic bias either in the research or the publication process (as we've mentioned, two very big ifs), then hopefully the aggregated result will give you a good idea of what the true effect is.

This is how science advances, at least in theory. Each time a new study comes out, it gets added to the pile; it's a new set of data points which – hopefully, on average – will bring the consensus of scientific understanding closer to the underlying reality.

But now imagine that a new study comes out, and instead of saying, 'This study adds to, and perhaps slightly shifts, our understanding of the underlying reality,' scientists immediately throw out all of the previous existing studies and say: 'This new study proves all the previous studies wrong: fish fingers now *cause* snoring, forget everything we said before.'

That is what goes on every time a journalist writes a news story about a new research paper – 'fish fingers cause snoring, groundbreaking new study reveals' – without putting it into the context of the existing research.

To be fair to journalists, this is a hard problem to solve. Newspapers report on news; in science, the most obvious 'news' is the publication of new studies. 'New study doesn't tell us very much and should only be viewed in the context of existing studies' is not a grabby headline, and besides, most journalists – just like most readers – may not realise that scientific papers need to be treated as part of a body, and don't stand alone, so they think, 'Oh I see this week red wine is good for us,' or whatever. Added to that, the increasingly straitened financial situation of many outlets means that science reporters are often writing five or more stories a day; they may well simply not have time to do more than write up the press releases, let alone phone up other scientists to put the new study in context.

It is, however, a problem, because it can give readers a misleading picture, both of the risk of specific things and of the scientific process itself. If it looks like the link between fish fingers and snoring changes every week – each time a new study comes out – then readers will be justified in thinking that science is basically making things up as it goes along.

Our silly thought experiment about fish fingers and snoring is one thing. But it goes on all the time, with real things. To continue picking on the *Daily Mail*, a Google search on their website for the exact phrase 'new study says' returns more than 5,000 results, on subjects ranging from the impact of obesity on brain function, to social media and stress, to whether coffee makes you live longer. Are those studies real? Yes. Does each one accurately portray the best current understanding of the science? Possibly not.

And it gets more serious. A study finding high levels of aluminium in the brains of autistic people[5] got some press attention back in 2017.[6] This study is not representative of the wider literature, which struggles to find any strong environmental effects on autism, but it added (because some vaccines contain aluminium) to the wider scares about vaccination.

Staying with vaccine scares and autism, the grandfather of them all – the 1998 study published in *The Lancet* by Andrew Wakefield et al., which apparently found a link between the MMR vaccine and autism – was itself very much an outlier.[7] A single, small study finding an unexpected result: a grown-up approach to science reporting would have considered it only of mild interest, even if it hadn't turned out to be fraudulent.[8] Instead, because of an industry-wide tendency to take single studies as the truth rather than as a single snapshot of a larger picture, it led to a gigantic health scare, a global drop in vaccination levels, and a small number of children ending up dead or disabled from measles.[9] Sometimes, just sometimes, giving an accurate impression of how important a single study is (usually: not very) really does matter.

What's the consensus on red wine and health, then? Well, despite the wildly varying headlines, the public-health position hasn't changed very much in years. People who drink small amounts of alcohol (roughly speaking, up to about seven pints of beer a week, or the equivalent) tend to live slightly longer than people who drink nothing at all; but as your alcohol consumption goes up above that, life expectancy drops off again. That finding has appeared over[10] and over[11] again[12] in large studies. It's described as a J-shaped curve: the mortality rate drops at first, then climbs up, in a shape like a slanting J or a Nike swoosh.

It's a small effect, and it's not *entirely* clear what's causing it – people who abstain from alcohol might, for instance, do so for health reasons, which might make them more likely to die prematurely. But the consensus does seem to be that there probably is a small protective effect from consuming small amounts of alcohol, compared to total abstention. Whether it's especially true of red wine is less clear.

But because the effect is small, any new study could easily find that a small amount of alcohol is bad for you, or good for you, or has no effect at all. It's only in context that new studies make sense. Be wary when you see something, especially something health-and-lifestyle-related, which includes the phrase 'new study says'.

Chapter 15

Demand for Novelty

'Does money make you mean?' asked a headline on *BBC News* from 2015.[1] It was discussing research into *money priming*, a psychological field which looks at how money affects our behaviour. Most dramatically, it talks about research finding that simply 'priming' someone with the idea of money, just by making them do sentence-unscrambling tasks with money-related words, makes them less likely to help others or donate to charity.[2]

Money priming, and the wider field it is part of, usually known as *social priming*, became popular in the first decade or so of the twenty-first century. It found remarkable results, like those discussed above, or – in the case of social priming – that priming someone with words related to age (like 'bingo', 'wrinkle' or 'Florida' – Americans associate Florida with retirement, apparently) makes them walk more slowly when they leave the experimenter's office.

Social priming was a *huge deal*. Daniel Kahneman, the great psychologist and pioneer of our understanding of cognitive biases – he won the Nobel Prize in Economic Sciences for his work with Amos Tversky – wrote in 2011 that 'disbelief is not an option' when it came to the astonishing priming effects.[3] A picture of a pair of eyes above an honesty box led people to put more money in it if than if there was a neutral picture of flowers.[4] Thinking about a shameful action, like stabbing a colleague in the back, leads people to buy more soap and disinfectant than they normally would, to scrub their soul clean: the 'Lady Macbeth effect.'[5]

But by the time that BBC article – and others, such as a long piece in *The Atlantic* from 2014[6] – were published, research into money priming was struggling. People were trying to find the same results as the early researchers and not finding them, or finding them to be much smaller and less impressive. What was going on?

Well, a lot of things. And there are many excellent books to read about the 'replication crisis' – the sudden realisation in many parts of science, but especially psychology, and *especially* social priming, that a huge swath of past research did not stand up to scrutiny. But the one we're going to look at is the demand for novelty in science.

There is a huge problem at the core of how science is done. It is not the fault of individual researchers exactly, although some do game the system. The problem exists also in how the popular media reports things – not just science, everything – although that is less surprising.

The problem is that *scientific journals want to publish scientific results that are interesting.*

That may not sound like a problem. You might think that publishing interesting results is exactly what scientific journals ought to be doing – after all, what's the point of publishing boring results that don't tell us anything new? But in fact it is a problem, a huge problem, and one that gets to the core of why many numbers that make it into news stories (and, perhaps more crucially, into the scientific literature) are wrong or misleading.

This demand for novelty is sometimes explicit. In 2011 a famous study rocked the world of psychological research: Daryl Bem's 'Feeling the future: Experimental evidence for anomalous retroactive influences on cognition and affect'.[7] That unwieldy title hid an apparently extraordinary finding: that people were psychic, clairvoyant. They could feel the future.

Bem's study took several classic psychological experimental methods and reversed them. One was a priming experiment, like the social priming mentioned above. Imagine that you want to find out whether you can change someone's behaviour with a subliminal image: a picture shown for a fraction of a second, too quick for your conscious mind to detect. You might show them two identical pictures of, say, a tree, one on the left of the screen, one on the right, and ask them to choose which they prefer. But just before the two images appear, an unsettling or unpleasant image – something violent or disgusting – pops into vision momentarily, either on the left or the right. Again, it's too quick to detect; but the hypothesis, which underpins the idea of 'subliminal advertising' which everyone got very excited about a decade or two ago, is that your unconscious mind would detect it. If it appeared on the left, you might be less likely to say you 'prefer' the tree on the left; if it appeared on the right, you might not choose the tree on the right. This was a common experimental model, part of the popular social priming subfield.

Bem's study did exactly that, but – in an interesting twist – it reversed the order. It made the priming image appear *after* the pictures of trees or whatever. And bizarrely, the subjects were *still* less likely to choose the tree that appeared in the same place as the nasty priming image. The effect was small, but statistically significant. This, the study suggested – entirely seriously – could only be the product of psychic ability.

Of course, having read this far, you will know that something else could explain it: simple fluke. Sometimes studies find false results just because the data is noisy. They might find the true result; they might find a greater, or smaller, result.

Most people reading this will probably think the 'true' level of psychic ability in the population is zero. But random errors in the data could, every so often, make a study return results that look as though it exists.

That's why science doesn't, or shouldn't, think in terms of single papers, as we saw in Chapter 14. Instead, it's concerned with where that study fits into the aggregate of all the other studies. You can get at that consensus position by doing meta-analyses and literature reviews: taking all the work on a topic and combining it. If one study finds that psychic powers are real, and ninety-nine find that they aren't, then you can probably write off the one outlier as a fluke.

For that to work, though, it is vital that all the studies performed on a subject are published. But – because scientific journals want to publish scientific results that are interesting – that doesn't happen. In the case of the Bem study, the way it didn't happen was obvious: a group of scientists, Stuart Ritchie, Richard Wiseman and Chris French, tried to replicate one of Bem's findings in a new study. They failed to do so; their experiment returned a null result. And the journal that published Bem's paper, the *Journal of Personality and Social Psychology*, refused to publish it.[8] It wasn't interested in boring replications of old work; it wanted novel new results.

In the end, the study did find a home, in the open-access journal *PLOS One*.[9] But if it hadn't, then someone trying to do a meta-analysis would have found one paper, Bem's, which found a result, and no others. The journal's demand for novelty would have led to an apparent scientific consensus that psychic powers are real. The Bem study, in fact, caused a huge ruction in psychology, because researchers realised that they had to accept one of two unpalatable truths: either psychic powers were real, or the experimental and statistical methods that underpin psychological science were capable of churning out meaningless nonsense.

(It's worth noting that Bem later *did* do a meta-analysis, which included Ritchie et al's paper and several others, and apparently *still* found that psychic abilities are real.[10] With checks for publication bias and everything. So either psychic powers are real, or the experimental and statistical methods that underpin psychological science are capable of churning out meaningless nonsense *even in meta-analyses*.)

This demand for novelty leads to a fundamental problem in science called *publication bias*. If 100 studies are carried out into whether psychic abilities are real and, say, ninety-two find that they're not and eight find that they are, that's a pretty good indicator that they're not. But if the journals, looking for novelty, only publish the eight that find a positive result, the world will be led to believe that we can see the future.

Silly psychic-powers studies are one thing, but if publication bias leads to doctors prescribing an exciting new cancer drug that doesn't in fact work, then that would be bad. Unfortunately, it happens. More than thirty years ago, the researcher R. J. Simes noted that published cancer studies which were registered in advance (registering studies in advance means they can't so easily be quietly filed away if they didn't find anything: see box for more details) were much less likely to return positive results than studies which weren't, suggesting that a lot of the unregistered studies were not being published.[11] A group reviewing the effectiveness of antidepressants found that thirteen out of fifty-five studies were simply never published; when the data from those studies was added back in, the apparent effectiveness of the antidepressants fell by a quarter.[12]

You do not need to read or understand this box, but if you would like to know about funnel plots and checking for publication bias, go ahead.

There's a clever way of checking whether there is publication bias in a field, known as a funnel plot. A funnel plot plots the results of all the studies on a topic, with smaller, weaker studies towards the bottom of the chart and larger, better studies towards the top. If there's no publication bias, then the studies should appear in a rough triangle shape: the smaller, less statistically powerful studies are widely spread out around the bottom (because you get more random error in a smaller study); the larger, more powerful studies are clustered more narrowly at the top. They should cluster around the same average, like this:

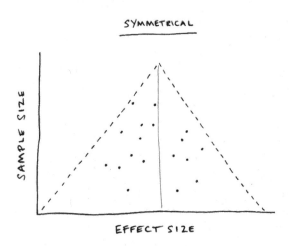

SYMMETRICAL

SAMPLE SIZE

EFFECT SIZE

But if some studies are being performed and then not published, you might not see that. Suddenly the studies which don't find anything don't appear. If instead of forming a neat triangle, you get a lopsided shape, like this:

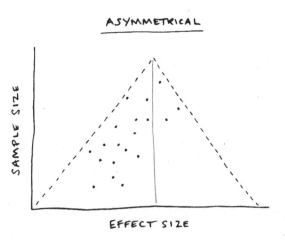

ASYMMETRICAL

SAMPLE SIZE

EFFECT SIZE

It's like a wall in a *Roadrunner* cartoon, covered in bullet holes, and you notice that in one area there's a Wile E. Coyote-shaped space where no bullets have hit. The absence itself is telling you that something is going on. Of course, it *could* be that just by chance the machine gun never happened to fire at that particular bit of wall. Or it could be that a luckless coyote was standing there, and the bullets never made it through him to hit the wall.

Similarly, it could be that just by chance a load of smaller, weaker studies happened to find bigger-than-average results, and none found any smaller-than-average ones. Or it could be that those studies *did* find those results and then – through the magic of publication bias – were never published, leaving a suspicious blank space where they ought to have been. There are other reasons that a funnel plot could look like this, but it's a hint that publication bias is a problem.

This isn't the only way of checking for publication bias – you can also simply write to researchers and ask them for any unpublished studies they might have performed, and then see whether the unpublished ones tend to return different results from the published ones. They often do.[13]

In the case of pharmaceutical companies, you could say that it is driven by naked corporate greed: if their antidepressant study finds the antidepressant doesn't work, then they won't be able to make so much money selling it. That might be part of the problem, although one study found that industry-sponsored trials were actually more likely to report their results within a year (as required by US law) than other ones.[14]

Instead, what's driving it, mainly, is that most journals select the things they publish on the basis of the results they find. So if you do a study on something – let's say whether humming 'La Marseillaise' before ordering at a restaurant makes people more likely to choose French fries – typically, you don't submit it to the journal when you come up with the idea; you submit it when your results are in.

'Humming "La Marseillaise" does not affect food choices' is an extremely boring title for a scientific journal, so most journals would reject it. But assuming that the humming really doesn't affect food choices, if twenty groups did the same study, then on average one of them would find a statistically significant ($p < 0.05$) result, just by fluke (assuming, as always, that the study was done properly). And that would get into the scientific literature, and then it would make headlines.

This is what happened with the money-priming studies mentioned at the top: a meta-analysis used funnel plots (see box) to determine whether publication bias was a problem, and found that it was.[15] Money priming may be a real effect, but it appears to be a lot smaller than it seemed in its heyday, because many of the studies which came back negative seem to have remained in researchers' file drawers.

It's worse than that, because scientists know that journals often won't publish negative results. So they won't even send them in. Or they will make little tweaks – perhaps reanalyse the data in a new way or remove some outliers – so that their results look positive. Scientific careers are 'publish or perish': if you're not getting papers published in scientific journals, then you won't progress, you won't get tenure. So scientists are hugely incentivised to get their papers published – in essence, they are incentivised to p-hack.

And it's worse than *that* if you're a reader of the popular press. Even if the studies do get published, the boring ones – the 'humming "La Marseillaise" does nothing of interest whatsoever' ones – won't get reported on. The demand for novelty is especially keen in the media, which is, after all, literally called 'the news'. Newspapers print stories about air crashes, which are novel and exciting and rare, rather than stories about planes that land safely, which is boring and common; so the public conversation, like the scientific literature, fills up with a skewed picture of how common exciting, dangerous things are. It's the same process.

There are noble steps to reduce this problem in science. The most promising are so-called 'Registered Reports' (RRs), in which journals agree to publishing studies on the basis of the methods, and then will do so regardless of the results, avoiding publication bias. One study compared standard psychological research to RRs in psychology and found that while 96 per cent of standard papers returned a positive result, only 44 per cent of RRs did, suggesting a huge problem.[16] RRs are catching on fast, and hopefully will become mainstream soon.

Of course, there's probably no realistic way to get the mainstream news media to report on boring studies that don't find anything, or on every aeroplane that successfully lands at Paris Charles de Gaulle. But the media *can* start making lots of noise about this problem in science, and hopefully the attention will get more journals on board with Registered Reports and other sensible reforms, because it is a fundamental problem and a huge reason why the numbers we read can't always be trusted.

Chapter 16

Cherry-picking

Let's go back in time to 2006, when Bob Carter, an Australian geologist, wrote a piece for the UK *Daily Telegraph*. 'There IS a problem with global warming,' it was headlined: 'it stopped in 1998.'[1] It was one of many articles written along those lines over the next eight years or so. The *Telegraph* was home to many of them, as was the *Mail on Sunday*.[2]

The idea that warming stopped in 1998 led to long discussion of the 'global-warming pause' or 'hiatus'. What explained this apparent slowdown (or, in some versions, reversal) of the temperature trend?

In all honesty, it's a complicated question, because the climate is a complicated thing. It's not a coincidence that when people think about chaos theory, they think about a butterfly flapping its wings in Brazil leading to a tornado in Texas. It's amazingly hard to predict and understand.

But, as it happens, there is a simple explanation. And the explanation is 'because you picked 1998 as your starting year'.

Imagine you're on a beach one afternoon. The waves come in and out; sometimes they reach higher up the beach, sometimes lower. You've built a sandcastle, and you're waiting for it to be destroyed by the tide. (This is a good thing to do with young children, to teach them about the remorselessness of time and the futility of all human endeavour.)

Foolishly, though, you didn't check before you left the holiday cottage whether the tide was coming in or going out. So every so often you look at how high the waves are reaching.

Most of the time, the waves are still falling a few feet short of the castle walls. Sometimes they're three feet short; sometimes two feet, sometimes four feet. But at, say, 3.50 p.m. one somewhat larger wave manages to get all the way up and splashes your crenellations. After that, they carry on reaching a little lower. If you record the highest wave in each five-minute section it might look like this, a wobbly but clear upward trend with one freakish outlier:

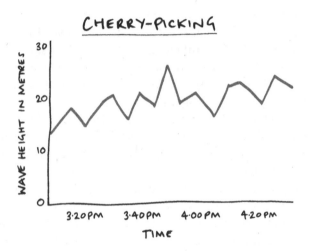

But now, say, you want to get everyone home, because it's the children's teatime. You need to convince your child that, actually, the tide isn't coming in, and there's no point waiting around to see the castle destroyed, so we may as well all bundle into the car. How can you do that?

Simple enough: you choose where to start your data series. You say to the child, 'Look, at 3.50 p.m. the tide reached 26 metres up the beach. But in the fifty minutes since then, it's never managed it again. There has been no rising of the tide since 3.50 p.m.'

That would be sort of true, in that the tide has never since reached the same zenith; but it would be misleading. If you'd picked any other time to measure from, you'd see a steady increase. That one freakishly large wave – which might have been caused by a passing speedboat, or frisky local whale – stands out from the data but it doesn't change the overall trend from low to high.

You'd have to be a weirdo to do that to your child, but people do it all the time with data. A *Sunday Times* front-page story in 2019 declared that the 'suicide rate among teenagers has nearly doubled in eight years', but the story had done the exact thing that our imaginary father did on the beach, albeit in reverse.[3] It measured from 2010, which was the lowest year for teen suicides in England and Wales on record.[4] Literally any year, measured from 2010, would have shown a rise (or you could pick any year *before* 2010, and show a decline).

Cherry-picking your start and end points is an example of what is known as 'hypothesising after results are known', or HARKing. It means getting your data, and then going through it to find exciting things. In noisy datasets, such as those on climate change or suicides, you'll find natural variation – things will go up and down for no particular reason, like the waves. If you want to, you can choose an unusually high or unusually low point to start or end on and use that to suggest that the trend is going up or down. You need to look more deeply than just at the highest and lowest to find longer-term trends, like the tides.

There are several other ways of HARKing. You can also select which bits of your data to look at, or the criteria you choose. For instance, the suicide story looked at teens – or specifically, fifteen- to nineteen-year-olds. None of the other groups showed a rise; and because, mercifully, suicides among teenagers are extremely rare, small random changes in the data can lead to large percentage swings. If you looked at young people in general, from ten- to twenty-nine-years-olds, there was no such jump.

Similarly with the climate data. Global surface air temperatures didn't reach 1998 levels for a long time – but as much heat energy is trapped in the top ten feet of the oceans as in the entire atmosphere.

This is a wider problem than simply climatology or media stories about suicide. HARKing, like the demand for novelty, is a huge problem in science. Oxford University's Centre for Evidence-Based Medicine found that papers published in the most highly respected medical journals in the world often change what they are looking for, after registering the trial, and without reporting that change in the paper.[5] That allows them to select start and end points, or even entirely different criteria for success. There may be good reasons to change what you're looking for, but it can also be a form of p-hacking, discussed in Chapter 5 (and you certainly ought to mention that you've done it).

This is often a hard problem to get around. You have to start your data series somewhere and, usually, it's going to be somewhere arbitrary. If the numbers jump around a lot, then it could make a big difference to the story your data appears to tell if you pick a low point or if you pick a high point. If, for instance, you are an incumbent government keen to show that you've improved child poverty, you might want to pick a starting year with particularly high child poverty and say, 'Look, it's fallen'; if you're the opposition party, you might want to pick a year with especially low poverty and say 'Look, it's risen'.

Taking in the wider picture helps, as does checking to see whether there are clear trends, or if it's just a noisy, wobbly line. But if you deliberately sift through the data to find the most dramatic places to start and finish your series, then you'll almost certainly end up telling a misleading story.

Incidentally, the papers largely stopped writing about the no-warming-since-1998 thing because 2014, 2015 and 2016 were all hotter – and all hotter than each other, a terrifying three 'hottest years on record' in a row. You can pick your outliers to tell your misleading stories if you like, but eventually the tide will come in anyway.

Chapter 17

Forecasting

Every few months, the UK's Office for Budget Responsibility issues forecasts for the country's economic performance, and – reasonably enough – the media reports those forecasts. For instance, in March 2019 the *Guardian* reported (accurately) that the OBR was predicting 1.2 per cent growth that year.[1] That was a relatively pessimistic estimate, but, the story said, the longer-term picture was more rosy.

Of course, the longer-term picture was not more rosy: almost exactly a year later, the UK went into lockdown to deal with Covid-19, and the economy shrank by 25 per cent in less than two months. Perhaps it's unfair to expect the OBR, or the *Guardian*, to have predicted a global pandemic. But when you see forecasts like that – 1.2 per cent growth forecast this financial year, unemployment to fall by 2 per cent this quarter, global temperatures to rise by 2.6° by 2100 – how are they being made? Should we trust them?

Let's forget about the economy for a moment, and start thinking about the weather in north London. The day we're writing this, the BBC weather app's Haringey location shows a symbol of a black cloud with rain coming out of it from 2 p.m. Upon seeing it, you might assume that that means it is going to rain at 2 p.m.

You would – probably – be wrong. Underneath the symbol is a percentage: 23 per cent. The weather app thinks that it's probably not going to rain at 2 p.m. – it thinks there's less than a one-in-four chance, in fact – but it's giving the 'rain' symbol anyway. (It was a bit more confident about later in the afternoon, if you're interested in the weather for a random day several months before you read this in a place where you probably don't live. By 7 p.m. it thought rain was 51 per cent likely. For what it's worth, at 2 p.m. it was still a beautiful clear blue sky.)

A weather forecast is not a mystical window into the future, a fortune-teller imparting wisdom: it is a best guess at a probability, intended to help you make decisions. You often hear about weather forecasts when they apparently get it wrong: the forecast gives you a big sunny symbol and says there's only a 5 per cent chance of rain; you plan a barbecue and invite all your friends; you just get the coals lit when suddenly the clouds scurry over, the heavens open and everyone's soaked to the bone still holding their undercooked beefburgers.

But it said a 5 per cent chance of rain, not 0 per cent. For every twenty times the app tells you there's a 5 per cent chance, you'd expect it to rain once. The odds of being dealt three of a kind in Texas Hold'em poker is about 5 per cent; if you've ever played it, you'll probably have got a few three-of-a-kind hands. If we were to play a hand right now, it's fairly unlikely that you'd get one, but if you play regularly, you wouldn't be surprised when one turns up. (Or if you play Dungeons & Dragons, you'll know how often you roll a 1 on a 20-sided die.)

Most likely, you won't remember the nineteen or so times when the weather app said it was a 5 per cent chance of rain and then it didn't rain. But you will remember the one time when it did.

This makes it hard to talk about whether the forecast was *right*. If the app tells you there's only a 1 per cent chance of rain, you might reasonably feel annoyed if you planned a day out and then it rained; but the forecasters could say, 'Well, we said there was a chance.' So how can you tell if a forecaster is any good or not? The odds of rain aren't mathematically determined, as they are with poker hands or D&D to-hit rolls.

It's simple: you look at a bunch of their forecasts, and you see whether the 1 per cent predictions come in 1 per cent of the time, and the 10 per cent predictions 10 per cent of the time, and so on. If they predicted a 5 per cent chance of rain 1,000 times, and it rained on about fifty of those times, then they were good at forecasting. If it rained much more than that – or less than that – then they were not. You can put numbers on how skilled a forecaster is.

As it happens, weather forecasting is – by the standards of most predictions of the future, anyway – extremely accurate: the UK Met Office, for instance, gets the next day's temperature right (to within 2°C) about 95 per cent of the time, and for three days' time 89 per cent of the time, according to its own blog in 2016.[2]

You do not need to read or understand this box, but if you would like to know more about how forecasting ability is judged, go ahead.

Forecasting skill can be measured with something called a Brier score. As we mentioned above, your score reflects how good at forecasting you are: if your 70 per cent guesses come in 70 per cent of the time, you're 'well calibrated'. But if your 70 per cent guesses come in 55 per cent of the time, you're over-confident, and if they come in 95 per cent of the time, you're under-confident.

But you don't just care about how well calibrated someone is. You also care about how specific they are. Saying that there's a 95 per cent chance of something happening – or a 5 per cent chance of it happening – is much more useful for making decisions than saying that there's a 55 per cent chance. If you're deciding whether to make a bet, support a policy or plan a barbecue, then someone who is both well calibrated and confident is of more use to you, as a forecaster, than someone who is well calibrated but vague.

Brier scores give you credit for being precise and right, but they punish you for being precise and wrong. They do that by using the *squared error*.

Say you make a forecast that it's 75 per cent likely that it will rain tomorrow. To get your Brier score, you divide your 75 by 100 to give you a number between 0 and 1 – in this case 0.75. Then you see whether it happened. If it did, you mark it as a 1; if it didn't, you mark it as 0.

The error is the difference between the outcome and how likely you said it would be. Let's say it did rain, so your outcome is 1. Your forecast was 0.75, so you subtract that from 1, then square the outcome (this is important, because it makes confident-but-right guesses score well, and confident-but-wrong ones score badly). It'll give you a score between 0 and 1, where 0 is perfect prediction and 1 is perfectly wrong – so the lower your score, the better, like in golf. In this case, your score is: $(1 - 0.75)2 = 0.0625$.

But if you'd got it wrong, you'd have landed in the 0.25 probability bit of your forecast, so the equation would look like this: $(1 - 0.25)2 = 0.5625$.

It can get a bit more complicated: often forecasters need to choose between several options, not just two, in which case the score is worked out in a slightly more complex way which gives an answer between 0 and 2. And things can get more complicated again in situations like temperature forecasts, which have many possible outcomes, rather than 'did rain' or 'didn't rain'. But it's the same fundamental system.

The Brier score was developed for weather forecasting, but it can be used for any clear, falsifiable predictions of the future. If you say it's 66 per cent likely that North Korea will have a new leader by this time next year, or that there's a 33 per cent chance that the Pittsburgh Steelers will win the 2021 Super Bowl, then those forecasts can be given Brier scores in exactly the same way as a weather forecast.

Sometimes you're not predicting something binary, like whether it will or won't rain. Sometimes you're predicting something that can vary – say the number of likely cases of malaria in Botswana in the next year, or (as in the examples we've mentioned earlier) GDP figures, or the temperature in Crouch End tomorrow. Then you don't want a simple yes or no, you want a number: the economy will grow by 3 per cent, say, or there will be 900 malaria cases.

Of course, it won't be exactly 3 per cent or exactly 900. You need to give an *uncertainty interval* – just as with p-values again, that's the area around your central prediction where you'd expect the real number to fall some percentage of the time (usually 95 per cent). So you might say that you predict that the temperature in Crouch End will be 18°C tomorrow, with a 95 per cent uncertainty interval of between 13°C and 23°C. The more confident a forecaster is, the narrower the uncertainty interval will be; but if they're very uncertain, then it will be very wide.

The weather is complicated – the paradigmatic example of a complex, chaotic system, in fact. But it is, in the end, physics; you can get better at understanding it with better algorithms and more powerful computers.

Weather isn't the only thing we try to forecast. We also try to predict the behaviour of humans – for instance economic growth, which is the behaviour of millions of humans across a country or a globe. And that is, in fact, more complicated, partly because humans respond to your predictions. If you say it's going to rain tomorrow, that's unlikely to affect whether it does or not. But if you predict that the stock market is going to go up, that might change whether people will buy shares.

Economists (one of us can wearily tell you) are often told that humans are too complicated to predict, so it's impossible to model them. But that isn't true; if it were, it would suggest that your guesses about their behaviour will be no better than chance, and that's clearly not the case. We can, for instance, predict with high confidence that you are not doing a headstand while you read this book; it is far more likely that you are sitting down. There are some fairly reliable predictions of human behaviour that you can make. And forecasts of the economy, or forecasts of elections based on opinion polling, do much better than random guessing would.

Forecasting is based on models. A forecast is a prediction: the economy will grow by 2 per cent, or there will be 12mm of rain over the weekend. A model is what you use to make that prediction, a simulation of a part of the world.

When we think of models, we think of complicated things like maths and equations. Often they are complicated, but they can also be simple.

Imagine you want to work out how likely it is that it will rain in the next hour. We're going to build a model, called the 'looking out of the window' model. The first thing we need to do, after looking out of the window, is to decide which pieces of information might reasonably help us.

An obvious one is the cloudiness of the sky. If the sky is a glorious blue with not a cloud in sight, then rain is very unlikely. If it's completely overcast, then rain is much more likely. If it's half and half, it's probably in between.

That gets you some of the way. Now you could add another bit of information: perhaps how dark the clouds are. We'd probably want to add lots more things – location, season, air temperature, wind speed. But let's start with just those two variables.

Writing out 'cloudiness times darkness of clouds equals probability of rain' every time would be a bit exhausting, so we use a shorthand. We could call cloudiness 'C', probability of rain 'R', and (just to give ourselves a bit of sciencey-sounding heft) the average cloud darkness 'β', the Greek letter beta. (It's our model, we can call it what we like.) The equation becomes: $\beta C = R$.

That equation is our model.

We look out of the window, and we see that the sky is overcast but that it's quite a light shade of grey – so 100 per cent cloud cover but 10 per cent on the greyscale chart. Those are your inputs; they go into your model. So 100 per cent times 10 per cent is 10 per cent, and our model puts out a 10 per cent chance of rain. That's your output.

Probably it will be terrible. What you need is feedback: you make predictions with your model, and you see how often it's right (does it rain as often as you say it will?), and you use them to update your model. Perhaps it turns out that the darkness of the cloud is more important, so you must give that more weight. Or perhaps not. But that's a model. You can make much more complicated ones – the Met Office's climate model contains over a million lines of code – but the principle is the same: data goes into the model and spits out an output.

Models of infectious disease, which became so prominent during the Covid-19 crisis, are another example: the classic example is the SIR model, which envisions the population as people who are either susceptible (S) to the disease, infected (I) or recovered and no longer susceptible (R). Essentially it treats them as dots which interact at random; given assumptions about how likely an infected person is to pass the disease on to a susceptible one, and how long it takes them to become infectious themselves, it can give you forecasts about how quickly the disease will spread among the real-world population. It can also be made more complicated, with more parameters such as people mixing in smaller groups or being differently susceptible, and you can add feedback from the real world, by comparing your forecasts to real outcomes and by looking at empirical data about how quickly people really spread the disease. Of course, your model is not the real world, so making it more complicated doesn't necessarily mean it will be more accurate; hence the need to look at how your model compares to actual outcomes.

Eventually, in some cases (such as weather), with experimentation and feedback, you can get fairly powerful, reliable predictions. But they're all uncertain. It's worth noting that in many cases, even 'forecasting' the present is hard: a majority of economists didn't think we were in recession even after the three most recent had begun.[3] Complicated things like economies are hard to understand.

*

So how about those financial forecasts? Well, in March 2019 the OBR did indeed, as we mentioned, forecast growth of around 1.2 per cent in 2020, and slightly faster growth thereafter. But it included a 95 per cent uncertainty interval of between about -0.8 per cent and 3.2 per cent in 2020.

The trouble is, the headline doesn't usually have room for 'growth will probably be somewhere between a fairly serious recession and a huge economic boom', so the central estimate, the 1.2 per cent, is usually all that gets reported.

(In this case, the real outcome will be well outside the 95 per cent uncertainty interval – a huge, double-digit fall in GDP. But that's probably OK, because devastating pandemics come along less than once every twenty years, so it shouldn't be inside your 95 per cent forecast.)

As a reader, you need to be aware of how forecasts are made, and you need to know that they are not mystical insights into fate – but nor are they random guesses. They're the outputs of statistical models, which can be more or less accurate; and the very precise numbers (1.2 per cent, 50,000 deaths, whatever) are central estimates inside a much bigger range of uncertainty.

The media, more importantly, has a duty to report that uncertainty, because being told 'the economy will grow by 1.2 per cent this year' might bring forth a very different response to being told 'the economy might shrink a bit or it might grow quite a lot or it might do anything in between, but our best guess is somewhere around 1.2 per cent growth.' We'd like the media to start treating readers and viewers like adults who can deal with uncertainty.

Chapter 18

Assumptions in Models

In late March 2020, the *Mail on Sunday* ran a piece by its entertainingly grumpy columnist Peter Hitchens, harrumphing about the models used to forecast the spread and death toll of Covid-19 in Britain and the world.[1] At the time there had been around 1,000 confirmed deaths in the UK,[2] but two weeks earlier, Imperial College London had released the findings of its model,[3] built by Professor Neil Ferguson and his team, which suggested that – if left unchecked – the toll could reach as high as 500,000. The government put the country into lockdown on the day of that model's release, 16 March.

By the time Hitchens wrote his piece, though, that estimate had changed. 'He has twice revised his terrifying prophecy, first to fewer than 20,000 and then on Friday to 5,700,' Hitchens wrote, saying that Ferguson was 'one of those largely responsible for the original panic'.

Is it true? Did the model really change its output that much? And is that evidence that the whole model was useless?

We spent the last chapter discussing modelling and how it works. But it's worth thinking a bit more about how they spit out the numbers they do. How does a model like Imperial's come up with a figure like half a million dead, while other models – such as that released by Oxford on 26 March – seemed to say wildly different things?[4] (And, if Hitchens is right, how come Imperial's own model seemed to say wildly different things a little while later?)

The answer comes down to the assumptions that those models

make. To look at those assumptions, we're going to talk about Brexit.

In the run-up to the June 2016 referendum, there were a lot of economic models flying around. Most of them predicted a negative impact on the economy;[5] one, very notably, predicted an economic boost. That was the model produced by Economists for Brexit, a group led by Patrick Minford; it suggested 'a welfare gain of 4% of GDP, with consumer prices falling 8%'.[6]

At the time of writing, we've only been out of the European Union for a few months; we are still in the transitional period, and so aligned with EU regulations and requirements. There is no way, so far, to establish who is correct; these models looked at the long-term impacts of Brexit and can only be judged in the long term.

That said, there were some models that made short-term forecasts which we *can* judge. The Treasury released the findings of its own economic model a few weeks ahead of the vote, and suggested that 'a vote to leave would represent an immediate and profound shock to our economy' which would 'push our economy into a recession', leaving GDP 3.6 per cent smaller and pushing 500,000 into unemployment.[7] That did not materialise. There was no recession.

What went wrong? Let's take a look at the different parts that influence GDP. Investment and manufacturing did fall, as the model suggested – driven by uncertainty around the UK's economic and trading future – but consumer spending remained high, and that kept the UK out of recession.

The modellers had assumed that consumer spending would fall. They had assumed this on the back of the 2008 financial crisis, in which consumer spending fell significantly, by more than £5 a week per person.[8] (For context, a fall of £5 is a big deal: every other year this millennium average consumer spending has gone up, apart from 2014–15, when it went down by 60p per week per person.)

Was that a bad assumption on the part of the modellers? With hindsight, clearly it was a *wrong* assumption; but that's easy to say now, and perhaps it was the reasonable best guess at the time.

Assumptions that modellers make are crucial to what ends up in their reports, and therefore in the media. The models are simply those assumptions, taken to their logical conclusions: if we assume that A=B and B=C, then our model tells us that A=C.

To some degree, that's what we're all doing all the time; we make all sorts of implicit assumptions when we make decisions. Written arguments, just as much as mathematical ones, rely on assumptions. Mathematical models have the advantage that many of those assumptions are made explicit: there's not much room for misinterpreting 'consumer spending will fall between 1 per cent and 5 per cent'.

The question is whether, and to what extent, the assumptions are *realistic*. But in and of themselves, unrealistic assumptions are not bad things. In the last chapter we made our own very silly model for weather forecasting. In it, we assumed that the greyness and extent of cloud cover would predict rain.

That sort of assumption is often grounded in empirical evidence: the Treasury forecast, for instance, was based on empirical evidence from behaviour following the financial crisis. And in the case of our grey-clouds assumption, we would try to base it on evidence by citing papers showing that overcast skies are associated with rain. (We are, however, not going to bother.)

But our model, being very basic, failed to include a lot of things. For instance, we didn't include any factor for location. Implicitly, therefore, it assumes that all places are identical; that the world is just a flat plain of identical landscapes. We know that's not true. That's not how reality is.

So we have an unrealistic assumption in our model. Does that mean our model is garbage?

Well, not necessarily. Adding location data might be able to improve your model's forecast, but it would be at the expense of adding more complexity: more data to gather, more computing power required. Whether that's worth it depends on how much extra accuracy the new details provide. It might not be that much of an issue when you're working with a silly model like ours, but when you're dealing with dozens of variables in a much larger, more complicated model, the accuracy/simplicity trade-off becomes very real. As statisticians say, 'The map is not the territory': your satnav doesn't have to tell you the colour of all the house doors to get you from A to B, but it does have to tell you where the road junctions are.

Sometimes you'll be happy to include the odd unrealistic assumption: many infectious-disease models (though not the Imperial College one), for instance, assume that everyone mixes at random. That's plainly not how humans interact; you're much more likely to bump into someone who lives on your street than you are someone who lives in a different city. But the model would be much more complex if it included those variables and might not gain all that much extra predictive power. It might be, for instance, that your basic model can predict the probability of rain to within a 10 per cent margin of error, and your more complex one can do it to within 5 per cent. Whether that's important depends on how accurate you need to be, and how much you have to pay – in terms of complexity and computing power – for that extra accuracy.

The problem is not when assumptions are unrealistic, necessarily, but when the unrealistic assumptions drastically affect the conclusions. Going back to the Economists for Brexit model, one reason it differed so much from the other forecasts was in its assumption about a concept known as 'economic gravity'. Under the law of physical gravity, how much two bodies interact depends on two things: how big they are, and how far apart they are. So the Earth's tides are very much affected by the moon, which is quite small but very close (on a cosmic scale); they're much less affected by Jupiter, which is very large, but rather further away.

Economic gravity works similarly. A nation's trade with another nation is affected by two things: how big that nation is, and how far away it is. So Britain trades more with France than it does with China, because France is medium-sized but close, while China is enormous but quite far away.[9] This is based on empirical observations[10] (it's the 'the most reliable empirical relationship in international economics', according to one critique of the Economists for Brexit model by LSE economists[11]) and forms a crucial assumption in most economic models.

The Economists for Brexit model, however, assumes that trade will happen equally well between countries that are far apart as countries that are close together, based entirely on how big the countries are and how cheap and high-quality the goods they produce are.

That's not – in the current global economy, at least – a realistic assumption. As we've seen, that in itself doesn't make it a bad one. It might be that you could get some quite accurate predictions by assuming that all countries trade equally regardless of distance, and that adding distance would add more complication to the model than would be helpful.

But it's the sort of assumption that can wildly change the output of the model, and it's important to understand the decision to include or not include it. The LSE critique found that using a model of trade which included the gravity equation changed the outcome from a 4 per cent boost to the economy to a blow 'equivalent to a 2.3% decline in UK income per capita', even if they held all of the Economists for Brexit model's other assumptions steady.

We're not here to declare a winner: it will be years before we know the impact of Brexit to any degree of certainty. And the partisan nature of the question means that those impacts will be highly disputed anyway, whether or not you have heard of the gravity equation.

So, what was going on with the Imperial model and its apparently changing output? Was Hitchens right to criticise it?

In short, not really. It's not that the Imperial model was above reproach, but that the criticisms Hitchens made were off target.

When he said that Ferguson had revised his model and was now predicting 5,700 deaths, he was simply confused: that was from a different model, by a different group of scientists (from Imperial's electrical engineering department, rather than epidemiology).[12] It was a far simpler model, sticking UK data on to the curve from China: by the time Hitchens' piece was published, one of the scientists who had produced that model had already revised his estimate up, to at least 20,000 dead.[13]

But how about the drop from 500,000 to 20,000? What went on there?

Well, the assumptions changed. One assumption in the model – probably several assumptions – involved people's behaviour, and how that affected the spread of the disease. Before lockdown happened, it was assumed that people would still be largely out and about, in contact with each other, and spreading the virus. After lockdown, it was assumed that they would be doing so much less. When they plugged that new assumption into the model, it gave a different number. In fact, the original 16 March paper modelled what would happen if there was something like a lockdown and predicted far fewer deaths than if there was no intervention.

It's worth remembering: if you're reading that 'a model' predicts something – a second wave of infections, an economic recession, 3°C of warming, a Tory victory in the next election – it really helps to know a bit about the assumptions that went into it. But often, that helpful detail gets lost in the news reports.

Chapter 19

Texas Sharpshooter Fallacy

Before the 2017 UK general election, the polling companies were almost uniformly confident that the Labour Party was going to get trounced. But ten days before the election, YouGov released a 'shock poll' (not, in fact, a poll, but a model of polls) that found that the Tories would lose about twenty seats, meaning that Theresa May, the prime minister, would lose her majority.[1]

Come election night, the results were revealed: the Tories lost thirteen seats, and YouGov's 'multilevel regression with post-stratification' (MRP) model had outperformed the rest of the field comfortably (with the final result well within their margin for error).[2]

Two and a half years later, with May gone and Boris Johnson in place, another election was called. This time, everyone had their eyes on the YouGov MRP, which called the election – in its final model, released days before the country voted – as a narrow Conservative win, with a majority of twenty-eight.[3] 'The new YouGov poll means this election is going to the wire,' reported one well-respected political journalist.[4]

The idea that we should have predicted things – the Covid-19 pandemic, the financial crisis, the outcome of the last election – is seductive. And when you find people who did predict something, it's tempting to believe that they had some extraordinary foresight, and that we should have listened to them at the time. But should we?

*

In 2019, a mobile-phone mast was moved in California. This might not seem like a big deal, but it made the news around the world.[5]

The mast was near a primary school in the Californian city of Ripon. It was moved after four children, all under the age of ten, were diagnosed with cancer. Getting cancer at that age is extremely rare.

But mobile phone masts do not cause cancer. (As good science communicators, we perhaps ought to say, 'there is no good evidence that mobile phone masts cause cancer,' but it has been pointed out to us that, to most people, 'there is no good evidence' sounds very similar to 'you can't prove anything, copper'. There is no epidemiological evidence for a link between phones and cancer, and there is no good theoretical reason to expect one, so we are happy to say that mobile phone masts do not cause cancer.)

So what happened with this cluster of cancer diagnoses?

Possibly something – there were suggestions of groundwater contamination, as well – but, equally possibly, nothing.[6] Every year, about 11,000 children under the age of fifteen are diagnosed with cancer in the USA.[7] The Ripon cases happened between 2016 and 2018, so you'd expect about 33,000 cases in that time. There are 89,000 elementary schools in the USA; a simple Poisson calculation (see box for what that is) suggests that about fifty of those schools would expect to see an outbreak of four or more in any given three-year period.

You do not need to read or understand this box, but if you would like to know how the Poisson distribution formula works, go ahead.

We wouldn't expect every school in the USA to have exactly the average number of cancer cases. There will be some random variation around the mean – it will be higher in some places and lower in others. That variation will look a bit like the normal distribution that we discussed in Chapter 3. But in order to determine how likely we are to see a given result in a specific time period, we need to look at something subtly different: the Poisson distribution.

In 1837, the French mathematician Siméon Denis Poisson published his *Researches into the Probabilities of Judgements in Criminal and Civil Cases*.[8] He was looking at how many wrongful convictions one would expect to see in French courts, given certain variables such as the number of jurors in a trial, the likelihood of any one to make a mistake and the prior probability of a suspect's guilt.

Something he needed to work out, in order to reach an answer, was: given on average something happens X times in a year (or an hour, or any given period), how likely is it that it will happen Y times in a year? On a graph, a Poisson distribution looks like this; the curve is made by joining the dots.

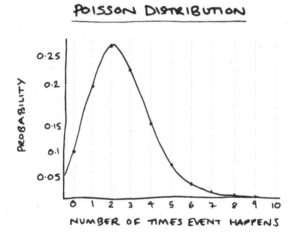

POISSON DISTRIBUTION

As the average gets lower, the curve gets taller and shifts to the left; as the average gets higher, the curve gets flatter and shifts to the right. The Y-axis gives the probability, up to a maximum of 1, and the X-axis gives the number of events. You look along the X-axis for the number of times you see the thing you're looking for, and you find the probability.

Let's say, for instance, you know that on average there are fifteen cases of cancer in a given school district each year; what are the chances that this year you see twenty? Plugging these numbers in (or just using a handy online calculator like we did) gives you 4 per cent, or 0.04.

But that's the probability of *exactly* twenty cases. You'd be just as startled to see twenty-one, or twenty-two. Instead you might want to know the chance of seeing twenty *or more* in any given year.

That might sound like it would take a very long time – that you'd have to work out the odds of twenty, then the odds of twenty-one, then the odds of twenty-two, and so on, all the way to infinity, and then add them together. But luckily, there's a shortcut.

We can take advantage of a concept called 'mutual exclusivity'. It means that some events cannot occur simultaneously: it is either one or the other. For example, if you roll a six, you can't have rolled a five, or a three. If you know that one of the outcomes *has* to happen, then the probabilities of the mutually exclusive events have to add up to one. If there's a one-in-six (0.167) chance of rolling a six, there must be a five-in-six (0.833) chance of not rolling a six; the odds of rolling *either* a six *or* not-a-six is 6/6, or 1.

So rather than working out the probability of seeing twenty or more cases of cancer, we can just work out the probability of *not* seeing them – of seeing between zero and nineteen. Then we can subtract this from one. So in our case, that would be the probability that there would be nineteen or less (nineteen, eighteen, seventeen and so on). We'd write that as $Pr(X<19)=0.875$. So $1- Pr(X<19)= P(X\geq20) \approx 0.125$, or 12.5 per cent.

There's a statistical mistake called the 'Texas sharpshooter fallacy'. The idea is that if someone shoots a machine gun randomly at a barn door, and then goes and draws a bullseye around any clusters of bullet holes they make, then they can make themselves look like a good shot. By comparison, if you take a random pattern of cancer cases spread across a country (or, since the story went global, across the world), and then draw a circle around random clusters, you can make it look like there's something going on, when there might not be.

This doesn't just apply to cancer clusters. It applies to people's predictions about the future as well. In 2008, with the global financial system on its knees, Her Majesty the Queen asked the question we were all asking: why didn't we see it coming? (Her actual quote, reported by an LSE economist, was: 'If these things were so large, how come everyone missed them?'[9]) It's a fair question, and one which economists and historians have been arguing over for the decade and a bit since.

But some people did see it coming, apparently. One such, it seems, was Vince Cable, who in 2008 was the Liberal Democrat Treasury spokesman. He had said in Parliament in 2003 that 'the growth of the British economy is sustained by consumer spending pinned against record levels of personal debt'[10] and that this, with manufacturing, exports and investment all faltering, would lead to disaster. He would be described as the 'sage of the credit crunch' by one newspaper, which added: 'if Mr Cable cannot see through the financial fog then no one can, or so the legend goes.'[11] This is a book about numbers, so it's worth pointing out that this is a fundamentally numerical prediction: Cable predicted that some numbers (particularly the numbers in the 'credit' columns of various major banks) were about to go downwards very rapidly.

Was he truly a sage? There's an old joke by the economist Paul Samuelson, who once said that the stock market 'has predicted nine of the last five recessions'.[12] Critics have suggested that Cable had done the same thing.[13] He made the prediction in 2003 (and apparently again in 2006); the crash didn't happen until 2008. He predicted another crash in 2017, but nothing particularly noticeable happened.[14] More relevantly, thousands of other MPs, journalists, academics and so on issued various pronouncements about what would and wouldn't happen to the economy in the coming years; some of them were always going to say the right thing. It's very unlikely that *you* will win the lottery, but someone's probably going to, and they didn't need any especial foresight to do so.

As we saw in Chapter 17, predicting the future is hard. Predicting the economy is even harder: if you can do it effectively, there's a good chance you're a billionaire. Being able to predict nine out of the five recessions – i.e. only being wrong four times – would in fact be an astonishingly good return.

But if you just go back and pick out people who *did* predict it, then there's a good chance you're falling for the Texas sharpshooter fallacy – you're taking a random scattering of data points and drawing a circle around the ones that happened to match the outcome.

It's not just journalists who do this. A 1993 study[15] apparently found links between power lines and childhood cancer in Sweden, and caused great interest,[16] even convincing the Swedish National Board for Industrial and Technological Development that electromagnetic radiation from power lines cause leukaemia in children. But statisticians pointed out that the study looked at 800 different health outcomes; the chances of a random cluster in one of them was very high.[17] (There is now, just as with mobile phones, no good reason to think that power lines cause cancer.)

The Texas sharpshooter fallacy can even put you in jail. A Dutch nurse, Lucia de Berk, spent six years in jail for murder after seven people died during her shifts, over three years. There was no forensic evidence that any of the deaths had been murder, let alone that they had been committed by her; but the cluster was suspicious enough for her to be convicted. As a statistician, Richard Gill, pointed out, this was a classic case of the Texas sharpshooter fallacy: sometimes people die on wards, and sometimes the same nurses will be present.[18] Ben Goldacre, in his *Guardian* column, noted that on one of Lucia de Berk's wards, there were six deaths in three years when she was supposedly killing people; in the three years before that, there were seven.[19] Her murders seemingly coincided with a sudden drop in the natural death rate. Randomness makes clusters, and if we draw circles around the clusters – bullseyes around the bullet holes – we can convince ourselves we're marksmen.

Remember the YouGov MRP poll? It had done extremely well in 2017, and everyone was interested in its prediction of a narrow Tory win in 2019.

But in the end, the result was a thumping landslide – an eighty-seat majority, with Labour collapsing in its northern heartlands. It's not that YouGov did conspicuously badly, but they didn't conspicuously outperform other polling companies either, many of whom predicted somewhat larger Tory majorities than the MRP model did. It may be that the MRP really did have some secret sauce in 2017 that allowed it to do better than the others; alternatively, it's perfectly possible that they all gave answers that were randomly scattered around an average, and the MRP happened to be the one that came closest. It's just not possible to tell from a single result.

If the MRP model consistently outperforms its rivals over the next several elections, then we can become more confident that it's doing something better. Otherwise, just as with the statistical significance problem we discussed in Chapter 5, we still can't reject the null hypothesis: that there's nothing there to explain.

Chapter 20

Survivorship Bias

How do you write a bestselling book? There's a formula,[1] apparently, or an algorithm, perhaps;[2] or it may be a secret code.[3] One article (the formula one) noted the success of J. K. Rowling, E. L. James and Alex Marwood, and suggested that being a woman with an androgynous pen name was a path to success. Another (the algorithm one) used text-mining software to find 2,800 common features of bestsellers: 'shorter sentences, voice-driven narratives and less erudite vocabulary', for instance; an 'emotional beat . . . an emotional high followed by a low, then another high, then another low'. Having authors who'd worked in journalism also helps, apparently, which is good news for us.

If your algorithm can predict with 97 per cent accuracy whether a book will become a bestseller, *just from the text alone*, you might think you'd try to use it to help write a few bestsellers of your own and make your millions before telling everyone else how you did it; but that's by the bye. The question we want to ask is: are these confident claims about how to write a bestseller based on anything real? Or have we bumped up against another statistical mistake?

Spoiler alert: it's a mistake. It's a very similar mistake to the Texas sharpshooter error discussed in the previous chapter, but with subtle and important distinctions. To understand it, let's talk about Second World War bombers, because that's fun.

In 1944, the US Navy was spending a lot of money, effort and lives in bombing Japanese runways. Their bombers were regularly shot at by enemy fighters and ground fire, and many were being destroyed. They wanted to reinforce their aircraft with armour plating; but armour plating is heavy, and you don't want to put it all over an aeroplane if you don't need to, because that will slow it down, make it less manoeuvrable and reduce its range and maximum payload.

So, reasonably enough, they looked at where returning planes had been damaged. They noticed that the bullet and flak-shrapnel impacts were mainly seen on the wings and fuselage, and not the engines. They decided that they ought to reinforce the wings and fuselage with extra armour plating.

A statistician, Abraham Wald, pointed out the problem with this.[4] The Navy was looking at a particular subset of planes – those planes *which had returned to the carrier*. The planes which had been hit a lot on the fuselage and wings tended to have made it back to base successfully. Those that had been hit on the engines, meanwhile, had predominantly fallen into the sea and not been counted in the statistics.

Without realising it, the US Navy had been basing its decision on a biased sample, such as we discussed in Chapter 4. This particular kind of sample bias is known as *survivorship bias*. It involves looking only at those members of a class which you get to hear about.

Douglas SBD Dauntless dive-bombers crashing into the Pacific Ocean is a particularly dramatic example, but there are many, more prosaic, examples of survivorship bias. The most obvious perhaps is 'secrets-of-my-success'-type books by business leaders. You know the kind of thing – *The 12 Habits of Extremely Rich People: How I Made My Millions by Getting Up Very Early, Only Drinking Avocado Smoothies and Firing 10 Per Cent of My Workforce at Random Every Two Weeks* by Thaddeus T. Richman.

We all want to know how to make our millions, so these books often sell well. But they are, usually, simply listing examples of survivorship bias.

SURVIVORSHIP BIAS

In his book *Standard Deviations*, the economist Gary Smith looked at two books which examined, between them, fifty-four well-performing companies, and which picked out common features about those companies – their corporate culture, dress code, whatever.[5] Smith noted that while the companies had indeed all outperformed the market up until the books were written, in the years since their publication almost exactly half of them had begun to underperform compared to the stock market – that is, do worse than the average company. The books which touted their excellent corporate cultures were looking at planes that landed, and noticing where the flak damage was, without wondering about what had happened to all the planes that never made it home.

There are other more everyday examples. The American mathematician Jordan Ellenberg describes the 'parable of the Baltimore stockbroker'.[6] One morning you receive a letter from an investment fund. It says: 'You should invest with us, because we always pick good stocks. But you won't believe us, so here's a free investment tip: buy Whoever Incorporated.' The next day, Whoever Incorporated stock goes up.

The next day, they send you another letter. 'Today, you should sell Thingummyjig Holdings.' The next day, Thingummyjig Holdings stock goes down.

They do this every day for ten days, and every time they get it right. On the eleventh day, they say: 'Now do you believe us? Would you like to invest?' They've picked ten in a row, so you think, yes! I can't lose! And you invest your children's university funds.

But what they've done is send out 10,000 letters; 5,000 of them say 'buy Whoever Incorporated', and 5,000 say 'sell'. If Whoever goes up, then the next day they send letters to the people who were told to buy; 2,500 of them say 'buy Thingummyjig' and 2,500 say 'sell'.

Then when Thingummyjig goes down, they send letters to the 2,500 people. And so on. After ten rounds of this, there'll be about ten people who've had ten successful tips in a row. They then invest all their money in this miracle stock-picker, who of course runs off with their cash. Derren Brown, the TV illusionist, used this exact method to pick five winning horses and then persuade a young mother to invest her life savings on the sixth.[7]

These sorts of frauds probably don't happen – Jordan Ellenberg told us via Twitter that he didn't know of any real-life Baltimore Stockbroker examples – but they can happen by accident. There are thousands of investment funds. A few of them achieve incredible rates of return for a while, so they get attention and lots of investment. But is that because they're genuinely beating the market or because they've got lucky, and you haven't noticed all the other mutual funds that have quietly gone bust?

Picture it like this. If you get 1,296 people, all wearing different-coloured hats, to roll a die, about 216 of them will roll a six. If you get those 216 to roll it again, you'll get about thirty-six more. If you ask those thirty-six to roll it again, you'll get about six sixes. Do it again, and you'll probably get one. Then you see the colour of the hat of the person who rolled the four sixes in a row and say: 'The secret to rolling four sixes in a row is wearing an orange-and-black-striped hat.' But it's easy to look back at what happens to have correlated with success in the past; what you need is to find what predicts success in the future. There is no reason to think that the orange-and-black-striped hat person will roll a six on their next go.

Survivorship bias is an example of a wider problem called *selecting on the dependent variable*. That sounds complex, but it's a simple idea: you can't work out why X happens if you only look at examples where X *did* happen. In a scientific experiment, the 'independent variable' is the thing you're changing (the dose of the drug you give them, say). The 'dependent variable' is the thing you're measuring to see if it changes (the patients' survival rates, perhaps).

So, imagine you want to know whether drinking water causes arthritis ('gets arthritis' is your dependent variable). If you look at all the people who developed arthritis, you'll quickly see that they all drank water. But because you didn't look at all the people who *didn't* develop arthritis, you have no idea whether arthritis patients tend to drink more water than the rest of us.

It might seem too obvious to mention, but it goes on all the time. Whenever there's a mass shooting, the media looks at the shooter,[8] and finds that they played violent video games.[9] Donald Trump did the same thing after the El Paso, Texas and Dayton, Ohio shootings in 2019.[10]

But this is as clear-cut an example of selecting on the dependent variable as the water-arthritis thing. The question is not 'do mass shooters play violent video games?', but 'do mass shooters play violent video games *more than anybody else?*'. (And then you have to ask about the causal arrow: do they become violent because they play the games, or do they play the games because they like violence? See Chapter 8 for a discussion of causality.)

Since a large majority of young men play violent video games, and almost all mass shooters are young men, it is extremely likely that any given mass shooter will have played *Call of Duty* or some other first-person shooter in the past. Saying that a mass shooter played a violent video game is only fractionally more surprising than saying that they ate pasta, or that they wore T-shirts. At least one study finds that, when you account for this, playing violent video games is associated with a *decline* in homicide rates, perhaps simply because the young men who would otherwise be out, and perhaps being violent, are at home playing *Grand Theft Auto V*.[11]

We've been talking about the media, but the problems of survivorship bias and selecting on the dependent variable probably have their most profound impact upstream of the news. The media often reports on scientific studies, but – obviously – only those scientific studies that get published. The trouble is, the scientific studies that get published – and the ones that make it into the news – aren't the only planes that set out from the carrier; they're just the ones that returned to base.

For reasons we discussed in Chapter 15 on demand for novelty, the scientific studies that get published are usually the ones that find interesting results.

Say there's an antidepressant drug you're testing. It doesn't actually do anything, but you don't know that yet. If you run ten studies, especially if they're quite small, they might all find slightly different things: five might find no effect; three might find it actually makes things worse; two might find a small improvement. In reality, it doesn't work; but just by fluke, these different studies return different results.

But because the novel, interesting (and, if you're the manufacturer, lucrative) result is 'this drug works', the studies that find a result are more likely to be published in a journal – as you'll remember from Chapter 15. So it may be that the eight studies which found either no effect or a negative effect sit in a scientist's file drawer somewhere, and when someone else comes to do a review of the evidence, they might find that the only two published studies on this antidepressant say that it works. And *then* doctors might prescribe this entirely ineffective antidepressant, because it looks as though the scientific evidence supports it.

This really happens and leads to real problems which kill real people. One study found that 94 per cent of published trials into antidepressants found positive results; but that when they sought out the unpublished ones and took those into account, that figure dropped to 51 per cent.[12]

There's another layer of this bias, which is that if you're reading about a scientific study in the mainstream media, it will have been deemed to have been interesting enough to put in the newspapers. 'New study finds that burnt toast doesn't, in fact, cause cancer', or 'Facebook actually isn't rotting children's brains, says research' probably won't make many headlines. If you're seeing a scientific story in the paper, remember that it's already flown two combat missions and returned to base. That doesn't mean it's not true, but it does give you reason to be wary; you don't know how many other studies into the same thing were shot down.

So: can you predict bestsellers using an algorithm? Does having an androgynous pen name help women get published? Well, we don't know, because we don't know how many female authors with androgynous pen names *didn't* get published. And can an algorithm predict with 97 per cent accuracy whether a manuscript will be a bestseller? Almost certainly not, unless it looked at all the books which didn't get into the bestseller lists, or which weren't published at all. You can look at all the mass shooters and see they played violent video games, but that tells you nothing about whether violent video games cause shootings; similarly, you can see that bestsellers share some features of vocabulary or plot, but that tells you nothing about whether those features helped sell the book. You're just looking at the planes that have made it back to base and pointing out all the bullet holes in their wings.

Chapter 21

Collider Bias

There was a strange phenomenon going on early in the Covid-19 pandemic. It was noticed that people who were hospitalised with the disease seemed to be less likely to smoke than the rest of the population.[1] The *Daily Mail* was one of the places that picked this up, and mentioned that French hospitals were even going to try giving nicotine patches to Covid patients.[2]

That is *really weird*. Smoking is staggeringly bad for you; probably the most directly dangerous thing that any large fraction of the population does. And it is dangerous because it plays hell with your respiratory system; it causes lung cancers, chronic obstructive pulmonary disease, emphysema, all these things that are bad for your ability to take a breath and get oxygen to the places it's meant to go. Since Covid-19 is a respiratory condition, you would expect smoking to make your survival chances worse, not better.

But however weird and counterintuitive the finding was, it kept turning up. What was going on?

There's a statistical anomaly that crops up from time to time, known as collider bias. It throws up some strange outcomes, making real relationships apparently disappear, or creating imaginary relationships out of nothing: it can even make things look like the opposite of reality.

We spoke in Chapter 7 about controlling for confounding variables. Let's imagine that you're conducting a study looking at how fast people can run. You notice something: on average, the more grey hairs a person has, the slower their time for the mile.

It could be that grey hairs slow you down. Or, perhaps more likely, it could be that both factors are linked to some third factor – maybe age. Perhaps getting older gives you grey hairs, and at the same time makes you run more slowly.

If you 'control' for how old someone is, you might find that the link disappears. Confounders like this can bias your results: unless you control for them, it can make your results seem over- (or under-) stated. This can create spurious links, like grey hair making you slower.

You can draw this on a diagram called a 'directed acyclic graph', showing how the causal arrows point: a 'confounder' is something that causes both the 'independent variable' which you're selecting on (grey hair), and the 'dependent variable' you think it might influence (running speed). We're interested in whether grey hairs influence running speed – the black arrow in the diagram below. But even though they're correlated, both are in fact influenced by a third factor, age, as shown by the white arrows.

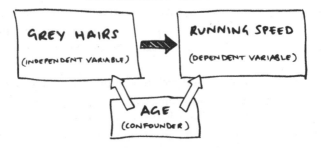

Controlling for confounding variables is necessary, and good statistical practice. But that doesn't mean you should just control for as many variables as you can, assuming they're all confounders: sometimes that can go wrong. Sometimes adding an extra variable to your analysis can make two things look like they're connected when they're not.

Here's an example. Suppose that acting talent and physical attractiveness are unrelated; if you're good at acting, you're no more (or less) likely to be beautiful than anyone else. The one tells you nothing about the other.

ACTING ABILITY VS. ATTRACTIVENESS

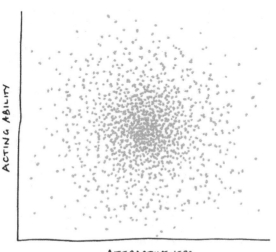

ATTRACTIVENESS

But now imagine there's a career path for people who are good at acting and/or beautiful, such as being a famous Hollywood actor. You're probably not going to become a famous actor if you're ugly *and* talentless, so most famous actors will be either one or the other or both.

Now, though, if you look at Hollywood actors and *only* Hollywood actors, you'll notice something. The most attractive ones tend to be less talented than the less attractive ones, even though – in the population at large – the two characteristics are unrelated.

That's because famous actors are selected on those two characteristics. If you're spectacularly attractive, you don't need to be all that good at acting, and vice versa. So all the unattractive bad actors are removed from the selection immediately, leaving a graph looking like this:

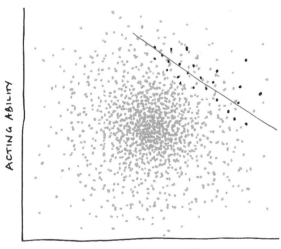

ACTING ABILITY VS. ATTRACTIVENESS

ACTING ABILITY

ATTRACTIVENESS

The same thing happens in US college admissions. You can get into college by being clever, or by being good at sports. In the general population, those two characteristics are unrelated, or only loosely related. But because you only need to be good at one or the other to get admitted, *within the population of US college students* sporting talent is negatively associated with academic ability. (Hence the 'dumb jock' stereotype.)

The examples we've used have been caused by the data you select – whether you just look at Hollywood actors, or American college students. But exactly the same would happen if you looked at all the data and then 'controlled for' these variables. For instance, children who get a fever might have food poisoning, or they might have influenza. (Or several other things, but let's stick with those two.) Let's say the two diseases are completely uncorrelated: there's no reason why if you've got one, you'd be more likely to have the other.

But if you were to conduct a study looking at the link between food poisoning and influenza, and you controlled for whether or not people have a fever, it could look as though children who had food poisoning are less likely to have the flu – that food poisoning protects against influenza somehow.

It's similar to the attractive-or-talented-but-rarely-both actors: it might be that if you've got a fever, you've probably got either food poisoning or the flu, but probably not both. But in this case, the bias isn't caused by only looking at a specific group (Hollywood actors). Instead, it's caused by the research-er thinking they are controlling for a confounding variable, to reduce bias – but in fact they've *added in* a collider variable and accidentally *created* bias.

A collider like this is the *opposite* of a confounder: where a confounder causes both the variables you're looking at, the two things you're looking at both cause a collider. So where controlling for a confounder helps reduce bias, controlling for (or selecting on) a collider can introduce it. (It's called a collider because the arrows 'collide'.) We can show it on a directed acyclic graph again – remember, the black arrow is what we're trying to investigate, and the white arrows show what influences what:

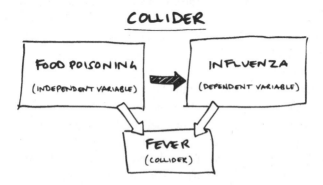

Real-life examples of collider bias in healthcare were first confirmed in 1978 and have been shown several times since.[3]

Is something like this going on in Covid-19 and smoking?
Possibly. A preprint published in May 2020 looked at the various
ways in which collider bias could be skewing our understanding of the Covid outbreak.[4] It noted that while there was a lot
of observational data coming in, the patients who were being
observed were not always representative of the wider population
– that they were being selected on quite specific grounds.

In the case of smoking, it pointed out that the sort of people
who got tested, early in the outbreak, were not random. Often,
they were healthcare workers. Healthcare workers tend to smoke
less than the population at large.

But the *other* sort of people who got tested a lot were people
with severe symptoms. So both 'being a healthcare worker' and
'having a severe case of Covid-19' cause people to have a Covid
test, and, if they get a *positive* result, be admitted to hospital.
But 'being a healthcare worker' is linked to non-smoking, so of
the population of people who had a positive Covid test, a large
percentage of them were non-smoking healthcare workers.

Remember the 'attractive or talented actors' example? This
is pretty much the same thing. Except instead of 'becoming a
famous actor', the thing we're selecting on is 'getting a positive
Covid test'. To have got a positive Covid test, you need to have
either 1) had obvious symptoms of Covid or 2) been a healthcare
worker (and therefore probably a non-smoker). If you have
neither of those things, you're not getting a test, so selecting on
having had a test could make the two appear correlated even if
they weren't.

The preprint demonstrated that even if there was no link
between smoking and Covid-19 severity at all, some realistic
assumptions about the prevalence of smoking in the population
at large versus the prevalence in the groups being tested could
create a strong apparent correlation. We can't at this stage be
sure that smoking doesn't protect against Covid, but since it's

counterintuitive, we need to treat it with great suspicion.

Spotting collider bias is difficult. For instance, some scientists claim that collider bias is behind the 'obesity paradox', the observed phenomenon that obese people seem to be less likely to die of diabetes than people of normal weight; others argue that it isn't.[5] It's a major ongoing controversy. If scientists struggle to establish what is and isn't collider bias, it's probably unfair to ask journalists and readers to do the same. But it's worth being aware that there are many ways in which correlations can be misleading, even when the study has done its best to control for other factors. Sometimes, controlling for them can even make the problem worse.

Chapter 22

Goodhart's Law

In April 2020, Britain – which was not at the time dealing *conspicuously* well with Covid-19 – was desperately trying to get its testing regime up and running.

It's hard to know exactly why some countries did well and some didn't; perhaps we'll learn more in future. But one noticeable thing was that a lot of countries which managed to suppress the spread of the disease early on seemed to have effective testing regimes. Britain, for a long time, was lagging behind.

So at the beginning of April, Matt Hancock, the government health secretary, pledged that by the last day of the month, Britain would be carrying out 100,000 tests every day.[1] At the time, it was managing somewhere around the 10,000 mark.[2]

It all became a bit strange after that. Political journalists, who are used to votes in Parliament or elections where crossing the magic threshold between 'not enough' and 'enough' really matters, started 'watching the numbers closely'.[3] The numbers were nowhere near enough by about 20 April. But then, in a weirdly *X-Factor*-ish moment, Hancock declared on live TV that on 1 May he could 'announce' (drum roll please) that 'the number of tests yesterday, on the last day of April, was 122,347.'[4] 'I knew that it was an audacious goal,' he added, 'but we needed an audacious goal, because testing is so important for getting Britain back on her feet.'

All's well that ends well, then? Well, not quite. It turned out that the bold 122,347 number concealed a multitude of problems.

For one thing, the original goal had been to *carry out* 100,000 tests a day. By the end of April, though, ministers were talking about having the 'capacity' to carry them out, and Hancock was emailing Conservative supporters pleading with them to sign up for testing.[5]

That was bad enough. But, more worryingly, the 122,347 figure included almost 40,000 tests which had only been put in the post, and not necessarily used.[6] Later in the month, as BBC Radio 4's *More or Less* programme would mercilessly document, it turned out that the government's testing figure included antibody tests given to determine whether someone had had the disease in the past, which are important things but totally different from the PCR test intended to tell whether someone has the disease now and therefore needs to isolate.[7] Also, the figure included people who had been tested multiple times on the same day because a first test had failed. The true number of diagnostic tests actually carried out on individuals was far lower: it was hard to determine, but well below the 100,000 number, and remained so well into May. The UK government ended up being reprimanded *twice* by its own statistical watchdog for playing fast and loose with the figures on the numbers of tests carried out.[8]

What went wrong, then? How could this simple number – the number of tests being carried out – end up getting so confused and misleading?

There's an old saying in economics, Goodhart's law, named for Charles Goodhart, a former economic adviser to the Bank of England: 'When a measure becomes a target, it ceases to be a good measure.' It may sound dry, but it has profound implications, and once you are aware of it you see it everywhere. What it means is that whatever metrics you use to assess how well you're doing at something, people will game those metrics.

The classic example is in education. Let's imagine that you notice that some pupils from some schools are doing better in life than pupils from other schools; they're getting into university more, ending up in work more, and generally flourishing and becoming well-rounded citizens. You look closer and you notice that the students at the flourishing schools are getting a higher percentage of A* to C grades at GCSE (or whatever) than the rest.

So you think, great. Here's a metric I can use to assess the performance of the schools. You start grading schools on what percentage of their pupils get those higher grades. The ones who get a higher percentage will be rewarded; the ones who get a lower percentage will be put into special measures, or their headteacher will be fired, or something.

Soon you notice that lots of schools start getting much higher percentages of A* to C grades. Even better! But you also notice that the children coming out of those schools do not seem to be the well-rounded examples of citizenry that you hoped for, despite their shiny certificates.

You can guess what happened. The teachers were under pressure from their bosses and educational authorities to get the A* to C numbers up. No doubt most valiantly tried their best, but found their promotion opportunities curtailed if they weren't hitting targets.

So some of them tried to find the quickest and easiest way of hitting that target. And the quickest and easiest way of doing that is not to give students a well-rounded education in the Aristotelian mode, ensuring a healthy mind in a healthy body, encouraging curiosity and drawing out the strengths of the individual. The quickest and easiest way is to give them hundreds of past exam papers and tell them what to expect. The quickest and easiest way is to game the system.

That was a hypothetical example, but something like it really happened: as the education researcher Daisy Christodoulou noted in 2013, when the number of A* to C grades was made a target in the UK, teachers started to game the target, notably by focusing disproportionate attention on pupils who were on the D/C borderline and therefore would do the most to push the numbers up.[9]

There is evidence in healthcare as well. In Oregon, hospital quality-of-care ratings are based on, among other things, in-hospital mortality: that is, what percentage of patients admitted go on to die. But doctors complained in 2017 that hospital administrators were refusing admission to some very sick patients on the basis that they were likely to die and would therefore bring the in-hospital mortality stats up.[10] In 2006 the US Medicare programme began the 'Hospital Readmissions Reduction Program' (HRRP), which measured how many heart-failure patients were readmitted to hospital within thirty days of being discharged. A 2018 study found that this actually increased mortality, apparently because hospitals delayed readmission to day thirty-one to avoid the patients appearing in their statistics.[11]

We've already discussed another example – the 'publish or perish' model in academia, where the worth of a scientist is measured by how many scientific papers they publish; and, relatedly, the convention which says that papers are a lot less likely to be published if they don't reach statistical significance (and find a positive result). It leads to scientists desperately trying to get papers published, even if they are worthless junk, and fiddling the statistics to get them to p<0.05, or simply filing them away if they find a null result. One study found that scientists were frequently gaming the metrics (things like the number of papers published, and the number of times each paper is cited), meaning that they are increasingly less useful as indicators of research quality.[12]

Business has the same problem. Tom, one of the authors of this book, is particularly aware that media companies which measure user engagement in terms of how many page views or unique users they receive tend to end up producing content that maximises those numbers, even at the expense of the quality of that content. (Memorably, an editor once insisted on having links on the home page click through to *another* home page, so that each reader would have to click twice to get to the actual story and would therefore be double-counted in the page-view statistics. Whatever your own line of work is, you can probably think of similar examples.)

The trouble is that these numbers are only a proxy for the thing we really care about. In the case of education, we care that schools produce well-rounded citizens who are prepared for adult life; we shouldn't really care about GCSEs per se (the fact that GCSE grades control access to later stages of education doesn't change that, it just makes the Goodhart's effect even stronger). We don't care how many patients are readmitted to hospital within thirty days, except insofar as that number tells us about the quality of care they received. And we don't care how many papers scientists publish or how widely they're cited, except insofar as they tell us about the quality of the science being done.

This isn't an argument against measuring things. You need to measure things in order to see whether they're going well: it's impossible, in a modern nation of millions of people, for government to individually assess every school and hospital. The same is true of any large modern business. Metrics are there for a reason: a car company might give a bonus to the salesperson who sells the most cars, for instance, and by incentivising them to work harder towards that goal could improve overall performance. Metrics are necessary.

But there's a trade-off. If your car salespeople start competing against each other, rather than co-operating – undermining each other in front of the customers – you could find that you sell fewer cars overall. If the people in charge aren't careful, they can lose sight of the fact that the metric isn't what you really care about, but is a proxy for an often complex, multifaceted and hard-to-define – but nonetheless real – underlying quality which you do care about. And people in the media can lose sight of it as well, so we get press releases about the number of 'items of PPE' produced, not caring about whether each 'item' is an N95 respirator mask or a single rubber glove.[13]

There are ways around Goodhart's law, to some degree: changing metrics frequently, or assessing things using multiple

metrics, can mitigate it. But no measurement will ever fully cap-
ture the underlying reality, which is always more complicated.
'Looking for the perfect summary statistic,' as the author Will
Kurt once tweeted, 'is like trying to write a dust jacket blurb that
replaces the need to read the book.'[14]

Obviously enough, that's what happened with the 100,000
tests target. (This isn't hindsight: Tom wrote ahead of the an-
nouncement that it was 'a hotbed for Goodhart's law'.[15]) The idea
– and it was a noble one – was to set a target which would, like
the bonus for car sales, drive up the number of tests. But then it
became vital to meet that particular arbitrary target, so suddenly
it was 'capacity to test', rather than tests carried out, and it was
tests in the post, and antibody tests.

The trouble was that we didn't care, really, whether precisely
100,000 tests were carried out; we cared about whether everyone
who needed a test could get one, and whether the testing regime
was fast and agile enough to inform people quickly that they had
Covid-19 and needed to self-isolate.

Whether the British response to Covid-19 – including testing – was adequate, and who is to blame for the ways in which
it wasn't, is a matter for the inevitable public inquiries that will
fill the next few years. But the idea that it mattered whether we
performed 99,999 or 100,001 tests on 30 April 2020 is ridiculous.
When reading about (or reporting on) targets, metrics and statistics, remember that they're proxies for the thing we care about,
not the thing itself.

Conclusion and Statistical Style Guide

One of us, Tom, has been a journalist for a distressing number of years now, and has worked at and for a range of different organisations. All of those organisations have something called 'house style'. It keeps the writing consistent. For instance, at the *Daily Telegraph*, where Tom started out, you would write '59 per cent', rather than '59%' (which our editor has also insisted on for this book). When naming someone for the first time in a piece, we would use their full name ('John Smith'), and at subsequent mentions an honorific ('Mr Smith') rather than just their surname. When writing about the pandemic, you'd say Covid-19, rather than COVID-19; when writing about the space agency, you'd say Nasa, not NASA.*

The *Telegraph* had its own style guide; one of its old columnists, Simon Heffer, turned it into a book.[1] It includes ways of referring to people and places; being the *Telegraph*, it was extremely particular about the correct ways of referring to members of the aristocracy, clergy and armed forces. ('Eldest sons of dukes, marquesses and earls take by courtesy the father's second title, assuming there is one. The Duke of Bedford's son is

---

* In case you're interested in that last rule: in general, UK newspapers prefer the former; US the latter. In the UK press, when you pronounce the initials as a single word – as in Covid, where you say ko-vid rather than See Oh Vee Eye Dee – you use the Title Caps format, only capitalising the first letter. But if you say the letters as letters – as in BBC, where you'd say Bee Bee See – then you cap them all. For some reason, this rule used to really upset supporters of the British political party Ukip.

the Marquess of Tavistock. Lord Tavistock's elder son, if he has one, can use the third title of the Duke, and he therefore is Lord Howland.')[2]

BuzzFeed, another outlet Tom worked at for some years, also has its own style guide, which is much less concerned with correctly identifying monsignors from baronets, and instead spends considerable time ruling on whether or not to hyphenate 'butt-dial', 'circle jerk' or 'douchebag' (all thus), or to put a space in 'J.Lo' when referring to the abbreviation for Jennifer Lopez (don't).[3] Other places have their own guides, and will place emphasis on the things that are most relevant to their readers.

(Memorably, the editor of the *Sunday Sport*, a particularly saucy UK tabloid, sent an email to all staff complaining about the headline 'MAN LOSES B*LLOCKS BUT DOCS SAVE HIS BELL-END!'. 'When I saw that page I felt physically sick,' he wrote. 'There are TWO glaring errors in the above headline that every member of editorial staff should spot straight away. Bollocks is NOT censored, even in headlines, and who the hell puts a hyphen in bellend?' He then listed a series of the 'commonest bungles' for staff to print off and keep on their desks: 'SHIT: Full out in copy and in headlines. WANK: Full out in copy, w**k in headlines . . .'[4])

Even smaller outlets which don't write their own style guides will have a house style: many US publications, for instance, will use the Associated Press Style Book.

All this is important: a well-applied house style helps make the writing consistent and clear, and makes the publication feel more professional. How can you trust anybody who writes 'bellend' in one paragraph and 'bell-end' in the next?

It is notable, though, that style guides rarely talk about how to present numbers. They talk about how to write them – newspapers usually write out one to nine in words (books usually go up to ninety-nine), larger numbers in numerals (134, or 5,299); then one billion, 10 billion. But they don't talk about how to use them carefully and responsibly; how to make sure the numbers themselves are telling a fair, accurate story.

This book can be read as just such a style guide: a sort of AP Style Book for statistical good practice. We hope that media outlets start to follow it, or (equally validly) see the need for one and then write their own. This is not just a book, in fact, but the start of a campaign for statistical literacy and responsibility in the media. If you're a journalist, we'd love you to start using it: if you're not a journalist, we'd love you to help us encourage media outlets to abide by it, or something similar.

Also, some of these tips might be quite useful as short reminders of what to be on the lookout for when you're reading stories yourself, whether you're a journalist or not.

We think this is necessary. It's not that you can't trust anything you read in the news; most journalists are decent people who want to write true stories. But they are, in Tom's experience, overwhelmingly words people, rather than numbers people. 'Data journalists' do exist, but they're specialists.

Most journalists are humanities graduates rather than STEM, at least according to the little data that we could find. This isn't a criticism – Tom studied philosophy at university, and you don't need a physics degree to gain the sort of basic numeracy required for journalism. But lots of journalists, like their readers, simply haven't had the occasion to think about how numbers ought to be presented.

You can't reasonably learn all there is to know about how to avoid statistical errors from one quite short book. And lots of the mistakes we've discussed are deep and systemic. For instance, avoiding Goodhart's law – the problem of measures becoming targets – is a huge issue at every level of government and business. The demand for novelty in science, let alone the media, is not something that we can do away with easily. Spotting collider bias, or Simpson's paradox, is difficult even for scientists, so it's not fair to blame journalists if they don't get it right.

But lots of the mistakes we've discussed in this book are easy to understand. If you haven't thought about them, then it might not occur to you to avoid them, but once they've been pointed out, almost anyone can see why they're an issue.

So, without further ado, here are our most important suggestions: our statistical style guide for numerically responsible journalists.

1) Put numbers into context

Ask yourself: is that a big number? If Britain dumps 6 million tons of sewage in the North Sea each year, that sounds pretty bad.[5] But is it a lot? What's the denominator? What numbers do you need to understand whether that is more or less than you'd expect? In this case, for instance, it's probably relevant that the North Sea contains 54 thousand billion tons of water. See Chapter 9 for more information.

2) Give absolute risk, not just relative

If you tell me that eating burnt toast will raise my risk of a hernia by 50 per cent, that sounds worrying. But unless you tell me how common hernias are, it's meaningless. Let readers know the absolute risk. The best way to do this is to use the expected number of people it will affect. For instance: 'Two people in every 10,000 will suffer a hernia in their lifetime. If they eat burnt toast regularly, that rises to three people in every 10,000.' And be wary of reporting on how 'fast-growing' something is: a political party, for instance, can easily be the 'fastest-growing' party in Britain if it has doubled in size from one member to two members. See Chapter 11 for more information.

3) Check whether the study you're reporting on is a fair representation of the literature

Not all scientific papers are born equal. When CERN found the Higgs boson, or LIGO detected gravitational waves, those findings were worth reporting on in their own right. But if you're reporting on a new study that finds that red wine is good for you, it should be presented in the context that there are lots of other studies, and that any individual study can only be part of the overall picture. Ringing up an expert in the field who *didn't* work on the study you're reporting on and asking them to talk you through the consensus on the subject is a good idea. See Chapter 14 for more information.

4) Give the sample size of the study – and be wary of small samples

A vaccine trial which has 10,000 subjects should be robust against statistical noise or random errors. A psychological study looking at fifteen undergraduates and asking whether washing their hands makes them feel less guilty is much less so. It's not that small studies are always bad, but they are more likely to find spurious results, so be wary of reporting on them; we somewhat arbitrarily suggest that if the study has less than 100 participants, it's a reason to be cautious. Some smaller studies can be very robust, and this is not a hard-and-fast rule; but, all else being equal, bigger is better. Relatedly, surveys and polls will often not have unbiased samples; be aware of them. See Chapter 3 for more information.

5) Be aware of problems that science is struggling with, like p-hacking and publication bias

Journalists can't be expected to be an expert in every field, and it's hard to blame them for missing problems in science that scientists themselves often miss. But there are some warning flags. For instance, if a study isn't 'preregistered', or better yet a Registered Report, then scientists might have gone back in once they've collected their data in order to find something that can get them a published paper. Alternatively, it might be that there are hundreds of other studies sitting unpublished in a scientist's desk drawer somewhere. Also, if a result is surprising – as in, it's

not what you'd expect given the rest of the findings in the field
– then that might be because it's not true. Sometimes science is
surprising, but most of the time, not very. See Chapters 5 and 15
for more information.

6) Don't report forecasts as single numbers. Give the confidence interval and explain it.

If you report that the Office for Budget Responsibility's model
says that the economy will grow 2.4 per cent next year, that sounds
accurate and scientific. But if you don't mention that their 95 per
cent uncertainty interval is between -1.1 per cent and +5.9 per
cent, then you've given a spurious sense of precision. The future
is uncertain, even though we sometimes wish it wasn't. Explain
how the forecast is made and why it's uncertain. See Chapters 17
and 18 for more information.

7) Be careful about saying or implying that something *causes* something else

Lots of studies find correlations between one thing and another – between drinking fizzy drinks and violence, for instance,
or between vaping and smoking weed. But the fact that two
things are correlated doesn't mean that one causes the other;
there could be something else going on. If the study isn't a
randomised experiment, then it's much more difficult to show
causality. Be wary of saying 'video games cause violence' or 'You-
Tube causes extremism' if the study can't show it. See Chapter 8
for more information.

8) Be wary of cherry-picking and random variation

If you notice that something has gone up by 50 per cent between 2010 and 2018, have a quick look – if you'd started your
graph from 2008 or 2006 instead, would the increase still have
looked as dramatic? Sometimes numbers jump around a bit, and
by picking a point where it happened to be low, you can make
random variation look like a shocking story. That's especially

true of relatively rare events, like murder or suicide. See Chapter 16 for more information.

9) Beware of rankings

Has Britain dropped from the world's fifth-largest economy to the seventh? Is a university ranked the forty-eighth best in the world? What does that *mean*? Depending on the underlying numbers, it could be a big deal or it could be irrelevant. For example, suppose that Denmark leads the world with 1,000 public defibrillators per million people, and the UK is seventeenth with 968. That isn't a huge difference, especially if you compare it with countries that have no public defibrillators. Does being seventeenth in this case mean that the UK health authorities have a callous disregard for public emergency first-aid installations? Probably not. When giving rankings, always explain the numbers underpinning them and how they're arrived at. See Chapter 13 for more information.

10) Always give your sources

This one is key. Link to, or include in your footnotes, the place you got your numbers from. The original place: the scientific study (the journal page, or the doi.org page), the Office for National Statistics bulletin, the YouGov poll. If you don't, you make it much harder for people to check the numbers for themselves.

11) If you get it wrong, admit it

Crucially – if you make a mistake and someone points it out, don't worry. It happens all the time. Just say thank you, correct it, and move on.

If you're an academic reading this book, there are things you can do to help as well. Just as with the media, you can't – on your own – be expected to fix all the structural problems in science, the publication bias and demand for novelty (although if you do make a point of using preregistration and Registered Reports, then bravo). But you can make sure that if your study has a press release, that press release accurately describes what is in it. Crucially, if your study *doesn't* show something, it's a good idea to explicitly say so. If, for instance, it finds that people who do crosswords are less likely to get Alzheimer's, but *doesn't* show a causal effect, it's worth saying in the press release: 'This does not mean that crosswords protect you against Alzheimer's.' Hearteningly, a study by a group of scientists at Cardiff University found in 2019 that these sorts of disclaimers in press releases reduced the amount of misinformation in media write-ups of the studies – but they *didn't* reduce the number of those write-ups.[6] Journalists were just as likely to write about the studies, but less likely to get the implications of their results wrong.

Of course, most of you (we hope) won't be journalists or academics, but ordinary, horny-handed peasants toiling the fields, or whatever it is normal people do. And we'd love it if you got involved too.

Trying to make these changes is a bit like trying to reform the voting system. In order to change to a new voting system – say from first past the post (FPTP), which the British Parliament uses, to proportional representation (PR), some form of which is used in many other European countries – you need to win under the old system. And once your political party has won under the old system, you have little incentive to change it, because you're in power.

Similarly, many academics and journalists know that there are problems with the way that numbers are presented. A lot of them openly acknowledge it. But once they're in a position of power – once they're professors or senior journalists – they've got there within the system, and they don't have much incentive to change it.

But if readers start demanding better – if they start writing to newspapers saying, 'Why haven't you given the absolute risk?' or 'Isn't that number cherry-picked?' – then the incentives change. By keeping an eye on the news and noting when they get it wrong in the ways we've described, and then politely pointing it out, you'll be helping improve the system, bit by bit. Or so we hope, anyway.

So, if you agree, we've set up a campaign: howtoreadnumbers. com. And that will make everyone better at statistics.

Well, maybe.

Credit: https://xkcd.com/552/

Acknowledgements

There are lots of people we would like to thank for helping us with this book. So, in no particular order:

Will Francis and Jenny Lord, agent and editor respectively, for being interested in the idea and helping us turn it from vague thought to actual saleable object.

Sarah Chivers, Tom's sister, who, conveniently, is a graphic designer, and who provided all the beautiful illustrations.

Pete Etchells, Stuart Ritchie, Stian Westlake, Mike Story, Jack Baker, Holger Wiese and someone apparently called 'Unlearning Economics', for examples, ideas and proof-reading.

Stephen, David's dad, for providing us with the example of Joe Wicks struggling to understand probability in Chapter 3. (Stephen did every session of 'PE with Joe' during lockdown.) And Andy, Tom's dad, for proof-reading.

Tom's children, Ada and Billy, for only occasionally bursting in and distracting him.

And, of course, our wives, Emma Chivers and Susanne Braun.

We would also like to make especial mention of Kevin McConway, Emeritus Professor of Applied Statistics at the Open University, who read the entire book, noted our many mistakes, and corrected them patiently and with great clarity and humour. No doubt some errors have made it through despite his super-human efforts, but that is our fault not his, and we would like to bow our heads in profound gratitude. Hail, King Kevin.

Notes

Chapter 1: How Numbers Can Mislead

1. Sam Blanchard, 'Care home "epidemic" means coronavirus is STILL infecting 20,000 people a day in Britain amid fears the virus's R rate may have gone back UP to 0.9', *Mail Online*, 2020 https://www.dailymail.co.uk/news/article-8297425/Coronavirus-infecting-20-000-people-day-Britain-warns-leading-expert.html

2. Nick McDermott, 'Coronavirus still infecting 20,000 people a day as spike in care home cases sends R rate up again to 0.9, experts warn', the *Sun*, 2020 https://www.thesun.co.uk/news/11575270/coronavirus-care-home-cases-spike/

3. Selvitella, A., 'The ubiquity of the Simpson's Paradox', *Journal of Statistical Distributions and Applications*, 4(2) (2017) https://doi.org/10.1186/s40488-017-0056-5

4. Simpson, Edward H., 'The interpretation of interaction in contingency tables', *Journal of the Royal Statistical Society*, Series B 13 (1951), pp. 238–41.

5. Persoskie, A. and Leyva, B., 'Blacks smoke less (and more) than whites: Simpson's Paradox in U.S. smoking rates, 2008 to 2012' *Journal of Health Care for the Poor and Underserved*, 26(3) (2015), pp. 951–6 doi:10.1353/hpu.2015.0085

Chapter 2: Anecdotal Evidence

1. Isabella Nikolic, 'Terminally-ill British mother, 40, who kept her lung cancer secret from her young daughter shocks medics after tumour shrinks by 75% following alternative treatment in Mexico', *Mail Online*, 2019 https://www.dailymail.co.uk/news/article-6842297/Terminally-ill-British-mother-40-shocks-medics-tumour-shrinks-75.html

2. Jasmine Kazlauskas, 'Terminally ill mum who hid cancer claims tumour shrunk 75% after "alternative care"', *Daily Mirror*, 2019 https://www.mirror.co.uk/news/uk-news/terminally-ill-mum-who-hid-14178690

3. Jane Lavender, 'I've cured my chronic back pain with £19 patch – but

NHS won't prescribe it', *Daily Mirror*, 2019 https://www.mirror.co.uk/news/
uk-news/ive-cured-chronic-back-pain-14985643

4. Hoy, D., March, L., Brooks, P., Blyth, F. et al., 'The global burden of low
back pain: estimates from the Global Burden of Disease 2010 study', *Annals
of the Rheumatic Diseases*, 73 (2014), pp. 968–74 https://ard.bmj.com/
content/73/6/968.abstract?sid=72849399-2667-40d1-ad63-926cf0d28c35

5. BioElectronics Corporation Clinical Evidence http://www.bielcorp.com/
Clinical-Evidence/

6. Andrade, R., Duarte, H., Pereira, R., Lopes, I., Pereira, H., Rocha, R. and
Espregueira-Mendes, J., 'Pulsed electromagnetic field therapy effectiveness
in low back pain: A systematic review of randomized controlled trials',
Porto Biomedical Journal, 1(5) (November 2016), pp. 156–63 doi:10.1016/j.
pbj.2016.09.001

7. Letter K192234, US Food & Drug Administration, 2020 https://www.
accessdata.fda.gov/cdrh_docs/pdf19/K192234.pdf

8. Stephen Matthews, 'Remarkable transformation of six psoriasis patients
who doctors say have been treated with homeopathy – including one
who took a remedy derived from the discharge of a man', *Mail Online*,
2019 https://www.dailymail.co.uk/health/article-7213993/Six-patients-
reportedly-cured-psoriasis-starting-homeopathy.html

Chapter 3: Sample Sizes

1. Ian Sample, 'Strong language: swearing makes you stronger,
psychologists confirm', the *Guardian*, 2017 https://www.theguardian.com/
science/2017/may/05/strong-language-swearing-makes-you-stronger-
psychologists-confirm

2. Stephens, R., Spierer, D. K. and Katehis, E., 'Effect of swearing on
strength and power performance', *Psychology of Sport and Exercise*, 35
(2018), pp. 111–17 https://doi.org/10.1016/j.psychsport.2017.11.014

3. 'PE with Joe, 2020 https://www.youtube.com/watch?v=H5Gmlq4Zdns

4. Gautret, P., Lagier, J.-C., Parola, P., Hoang, V. T. et al.,
'Hydroxychloroquine and azithromycin as a treatment of COVID-19:
Results of an open-label non-randomized clinical trial', *International
Journal of Antimicrobial Agents*, 56(1) (2020) doi:10.1016/j.
ijantimicag.2020.105949

5. Donald J. Trump, @realdonaldtrump, 2020 https://twitter.com/
realDonaldTrump/status/1241367239900778501

Chapter 4: Biased Samples

1. 'BRITS LIKE IT CHEESY: Cheese on toast has been voted the nation's
favourite snack', the *Sun*, 2020 https://www.thesun.co.uk/news/11495372/

brits-vote-cheese-toast-best-snack/

2. Bridie Pearson-Jones, 'Cheese on toast beats crisps and bacon butties to be named the UK's favourite lockdown snack as people turn to comfort food to ease their anxiety', *Mail Online*, 2020 https://www.dailymail.co.uk/femail/food/article-8260421/Cheese-toast-Britains-favourite-lockdown-snack.html

3. 'How do you compare to the average Brit in lockdown?', Raisin.co.uk, 2020 https://www.raisin.co.uk/newsroom/articles/britain-lock-down/

4. Greg Herriett, 20 November 2019, Twitter https://twitter.com/greg_herriett/status/1197115377739845633

5. Ofcom media literacy tracker 2018 https://journals.sagepub.com/doi/10.1177/2056305117698981

6. Sloan, L., 'Who Tweets in the United Kingdom? Profiling the Twitter population using the British Social Attitudes Survey 2015', *Social Media + Society*, 3(1) (2017) https://doi.org/10.1177/2056305117698981

7. Tversky, A. and Kahneman, D., 'The behavioral foundations of economic theory', *The Journal of Business*, 59(4) (October 1986), Part 2, pp. S251–S278.

Chapter 5: Statistical Significance

1. Helena Horton, 'Men eat more food when they are trying to impress women, study finds', the *Daily Telegraph*, 2015 https://www.telegraph.co.uk/news/science/12010316/men-eat-more-food-when-they-are-trying-to-impress-women.html

2. Lisa Rapaport, 'Men may eat more when women are watching', *Reuters*, 2015 https://www.reuters.com/article/us-health-psychology-men-overeating/men-may-eat-more-when-women-are-watching-idUSKBN0TF23120151126

3. 'Men eat more in the company of women', 2015, *Economic Times* https://economictimes.indiatimes.com/magazines/panache/men-eat-more-in-the-company-of-women/articleshow/49830582.cms

4. Kniffin, K. M., Sigirci, O. and Wansink, B., 'Eating heavily: Men eat more in the company of women', *Evolutionary Psychological Science*, 2 (2016), pp. 38–46 https://doi.org/10.1007/s40806-015-0035-3

5. Cassidy, S. A., Dimova, R., Giguère, B., Spence, J. R. and Stanley, D. J., 'Failing grade: 89% of introduction-to-psychology textbooks that define or explain statistical significance do so incorrectly', *Advances in Methods and Practices in Psychological Science*, 2(3) (2019), pp. 233–9 https://doi.org/10.1177/2515245919858072

6. Haller, H. and Kraus, S., 'Misinterpretations of significance: A problem students share with their teachers?', *Methods of Psychological Research*, 7(1) (2002), pp. 1–20.

7 Cassidy et al., 2019.

8. Brian Wansink, 'The grad student who never said "No"', 2016, archived at https://web.archive.org/web/20170312041524/http:/www.brianwansink.com/phd-advice/the-grad-student-who-never-said-no

9. Stephanie M. Lee, 'Here's how Cornell scientist Brian Wansink turned shoddy data into viral studies about how we eat', *BuzzFeed News*, 2018 https://www.buzzfeednews.com/article/stephaniemlee/brian-wansink-cornell-p-hacking

10. Ibid.

Chapter 6: Effect Size

1. Jean Twenge, 'Have smartphones destroyed a generation?', *The Atlantic*, 2017 https://www.theatlantic.com/magazine/archive/2017/09/has-the-smartphone-destroyed-a-generation/534198/

2. Dr Leonard Sax, 'How social media may harm boys and girls differently', *Psychology Today*, 2020 https://www.psychologytoday.com/us/blog/sax-sex/202005/how-social-media-harms-boys-and-girls-differently

3. Orben, A. and Przybylski, A., 'The association between adolescent well-being and digital technology use', *Nature Human Behaviour*, 3(2) (2019) doi:10.1038/s41562-018-0506-1

4. Damon Beres, 'Reading on a screen before bed might be killing you', *Huffington Post*, 23 December 2014 https://www.huffingtonpost.co.uk/entry/reading-before-bed_n_6372828

5. Chang, A. M., Aeschbach, D., Duffy, J. F. and Czeisler, C. A., 'Evening use of light-emitting eReaders negatively affects sleep, circadian timing, and next-morning alertness', *Proceedings of the National Academy of Sciences of the United States of America*, 112(4) (2015), pp. 1232–7 doi:10.1073/pnas.1418490112

6. Przybylski, A. K., 'Digital screen time and pediatric sleep: Evidence from a preregistered cohort study', *The Journal of Pediatrics*, 205 (2018), pp. 218–23.e1

Chapter 7: Confounders

1. Arman Azad, 'Vaping linked to marijuana use in young people, research says', *CNN*, 2019 https://edition.cnn.com/2019/08/12/health/e-cigarette-marijuana-young-people-study/index.html

2. Chadi, N., Schroeder, R., Jensen, J. W. and Levy, S., 'Association between electronic cigarette use and marijuana use among adolescents and young adults: A systematic review and meta-analysis', *JAMA Pediatrics*, 173(10) (2019), e192574 doi:10.1001/jamapediatrics.2019.2574

3. Hannah Ritchie and Max Roser, 'CO_2 and greenhouse gas emissions',

Our World in Data https://ourworldindata.org/co2-and-other-greenhouse-gas-emissions#per-capita-co2-emissions; Hannah Ritchie and Max Roser, 'Obesity', *Our World in Data* https://ourworldindata.org/obesity

4. Camenga, D. R., Kong, G., Cavallo, D. A., Liss, A. et al., 'Alternate tobacco product and drug use among adolescents who use electronic cigarettes, cigarettes only, and never smokers', *Journal of Adolescent Health*, 55(4) (2014), pp. 588–91 doi:10.1016/j.jadohealth.2014. 06.016

5. van den Bos, W. and Hertwig, R., 'Adolescents display distinctive tolerance to ambiguity and to uncertainty during risky decision making', *Scientific Reports 7*, 40962 (2017) https://doi.org/10.1038/srep40962

6. Zuckerman, M., Eysenck, S. and Eysenck, H. J., 'Sensation seeking in England and America: cross-cultural, age, and sex comparisons', *Journal of Consulting and Clinical Psychology*, 46(1) (1978), pp. 139–49 doi:10.1037//0022-006x.46.1.139

7. Dai, H., Catley, D., Richter, K. P., Goggin., K and Ellerbeck, E. F., 'Electronic cigarettes and future marijuana use: A longitudinal study', *Pediatrics*, 141(5) (2018), e20173787 doi:10.1542/peds.2017-3787

8. Sutfin, E. L., McCoy, T. P., Morrell, H. E. R., Hoeppner, B. B. and Wolfson, M., 'Electronic cigarette use by college students', *Drug Alcohol Depend*, 131(3) (2013), pp. 214–21 doi:10.1016/j.drugalcdep.2013.05.001

Chapter 8: Causality

1. 'Fizzy drinks make teenagers violent', the *Daily Telegraph*, 2011 https://www.telegraph.co.uk/news/health/news/8845778/Fizzy-drinks-make-teenagers-violent.html

2. 'Fizzy drinks make teenagers more violent, say researchers', *The Times*, 2011 https://www.thetimes.co.uk/article/fizzy-drinks-make-teenagers-more-violent-say-researchers-7d266cfz65x

3. Solnick, S. J. and Hemenway, D., 'The "Twinkie Defense": The relationship between carbonated non-diet soft drinks and violence perpetration among Boston high school students', *Injury Prevention*, 18(4) (2012), pp. 259–63.

4. Angrist, J. D., 'Lifetime earnings and the Vietnam era draft lottery: Evidence from Social Security administrative records', *The American Economic Review*, 80(3) (1990), pp. 313–36 www.jstor.org/stable/2006669

5. Haneef, R., Lazarus, C., Ravaud, P., Yavchitz, A. and Boutron, I., 'Interpretation of results of studies evaluating an intervention highlighted in Google health news: A cross-sectional study of news', *PLOS One*, 16(10) (2015), e0140889

6. Miguel, E., Satyanath, S. and Sergenti, E., 'Economic shocks and civil conflict: An instrumental variables approach', *Journal of Political Economy*,

112 (4) (2004), pp. 725–53 www.jstor.org/stable/10.1086/421174

7. Jed Friedman, 'Economy, conflict, and rain revisited', World Bank Blogs, 21 March 2012

8. Antonakis, J., Bendahan, S., Jacquart, P. and Lalive, R., 'On making causal claims: A review and recommendations', *The Leadership Quarterly*, 21 (2010), pp. 1086–1120 10.1016/j.leaqua.2010.10.010.

Chapter 9: Is That a Big Number?

1. '£350 million EU claim "a clear misuse of official statistics"', *Full Fact*, 2017 https://fullfact.org/europe/350-million-week-boris-johnson-statistics-authority-misuse/

2. Sir David Norgrove, letter to Boris Johnson, 17 September 2017 https://uksa.statisticsauthority.gov.uk/wp-content/uploads/2017/09/Letter-from-Sir-David-Norgrove-to-Foreign-Secretary.pdf

3. TFL Travel in London Report 11, 2018 http://content.tfl.gov.uk/travel-in-london-report-11.pdf

4. Rojas-Rueda, D., de Nazelle, A., Tainio, M. and Nieuwenhuijsen, M. J., 'The health risks and benefits of cycling in urban environments compared with car use: Health impact assessment study', *British Medical Journal*, 343 (2011), d4521–d4521.

5. Kaisha Langton, 'Deaths in police custody UK: How many people die in police custody? A breakdown', *Daily Express*, 2020 https://www.express.co.uk/news/uk/1292938/deaths-in-police-custody-uk-how-many-people-die-in-police-custody-UK-black-lives-matter

6. 'Police powers and procedures, England and Wales', year ending 31 March 2019, 24 October 2019 https://www.gov.uk/government/collections/police-powers-and-procedures-england-and-wales

7. Arturo Garcia and Bethania Palma, 'Have undocumented immigrants killed 63,000 American citizens since 9/11?', *Snopes*, 22 June 2018 https://www.snopes.com/fact-check/have-undocumented-killed-63000-us-9-11/

8. 'Crime in the US 2016', FBI, 25 September 2017 https://ucr.fbi.gov/crime-in-the-u.s/2016/crime-in-the-u.s.-2016/

9. Budget 2020 https://assets.publishing.service.gov.uk/government/uploads/system/uploads/attachment_data/file/871799/Budget_2020_Web_Accessible_Complete.pdf

Chapter 10: Bayes' Theorem

1. Zoe Zaczek, 'Controversial idea to give "immunity passports" to Australians who have recovered from coronavirus – making them exempt from tough social distancing laws', *Daily Mail Australia*, 2020 https://www.dailymail.co.uk/news/article-8205049/Controversial-idea-introduce-

COVID-19-immunity-passports-avoid-long-term-Australian-lockdown.
html

2. Kate Proctor, Ian Sample and Philip Oltermann, '"Immunity passports"
could speed up return to work after Covid-19', the *Guardian*, 2020 https://
www.theguardian.com/world/2020/mar/30/immunity-passports-could-
speed-up-return-to-work-after-covid-19

3. James X. Li, FDA, 1 April 2020 https://www.fda.gov/media/136622/
download

4. Nelson, H. D., Pappas, M., Cantor, A., Griffin, J., Daeges, M. and
Humphrey, L., 'Harms of breast cancer screening: Systematic review to
update the 2009 U.S. Preventive Services Task Force Recommendation'
Annals of Internal Medicine, 164(4) (2016), pp. 256–67 doi:10.7326/M15-
0970 (published correction appears in *Annals of Internal Medicine*, 169(10)
(2018), p. 740)

5. Brawer, M. K., Chetner, M. P., Beatie, J., Buchner, D. M., Vessella, R. L.
and Lange, P. H., 'Screening for prostatic carcinoma with prostate specific
antigen', *Journal of Urology*, 147(3) (1992), Part 2, pp. 841–5 doi:10.1016/
s0022-5347(17)37401-3

6. Navarrete, G., Correia, R., Sirota, M., Juanchich, M. and Huepe, D.,
'Doctor, what does my positive test mean? From Bayesian textbook tasks
to personalized risk communication', *Frontiers in Psychology*, 17 September
2015 doi:10.3389/fpsyg.2015.01327

7. Jowett, C., 'Lies, damned lies, and DNA statistics: DNA match testing,
Bayes' theorem, and the Criminal Courts', *Medicine, Science and the Law*,
41(3) (2001), pp. 194–205 doi:10.1177/002580240104100302

8. *The Times* Law Reports, 12 January 1994.

9. Hill, R., 'Multiple sudden infant deaths – coincidence or beyond
coincidence?', *Paediatric and Perinatal Epidemiology*, 18(5) (2004), pp.
320–26 doi:10.1111/j.1365-3016.2004.00560.x

10. Anderson, B. L., Williams, S. and Schulkin, J., 'Statistical literacy of
obstetrics-gynecology residents', *Journal of Graduate Medical Education*,
5(2) (2013), pp. 272–5 doi:10.4300/JGME-D-12-00161.1

Chapter 11: Absolute vs Relative Risk

1. Sarah Knapton, 'Health risk to babies of men over 45, major study warns',
the *Daily Telegraph*, 2018 https://www.telegraph.co.uk/science/2018/10/31/
older-fathers-put-health-child-partner-risk-delaying-parenthood/

2. Khandwala, Y. S., Baker, V. L., Shaw, G. M., Stevenson, D. K., Faber,
H. K., Lu, Y. and Eisenberg, M. L., 'Association of paternal age with
perinatal outcomes between 2007 and 2016 in the United States: Population
based cohort study', *British Medical Journal*, 363 (2018), k4372.

3. Sarah Boseley, 'Even moderate intake of red meat raises cancer risk,

study finds', the *Guardian*, 2019 https://www.theguardian.com/society/2019/
apr/17/even-moderate-intake-of-red-meat-raiscs-cancer-risk-study-finds
4. Ben Spencer, 'Teenage boys' babies are "30% more likely to develop
autism, schizophrenia and spina bifida"', the *Daily Mail*, 2015 https://www.
dailymail.co.uk/health/article-2957985/Birth-defects-likely-children-teen-
fathers.html
5. Bowel cancer risk, Cancer Research UK https://www.cancerresearchuk.
org/health-professional/cancer-statistics/statistics-by-cancer-type/bowel-
cancer/risk-factors
6. Klara, K., Kim, J. and Ross, J. S., 'Direct-to-consumer broadcast
advertisements for pharmaceuticals: Off-label promotion and adherence
to FDA guidelines', *Journal of General Internal Medicine*, 33 (2018), pp.
651–8 https://doi.org/10.1007/s11606-017-4274-9 https://link.springer.com/
article/10.1007/s11606-017-4274-9/tables/6
7. Kahwati, L., Carmody, D., Berkman, N., Sullivan, H. W., Aikin, K.
J. and DeFrank, J., 'Prescribers' knowledge and skills for interpreting
research results: A systematic review', *Journal of Continuing
Education in the Health Professions*, 37(2) (Spring 2017), pp. 129–36
doi:10.1097/CEH.0000000000000150 https://journals.lww.com/jcehp/
Abstract/2017/03720/Prescribers__Knowledge_and_Skills_for_
Interpreting.10.aspx

Chapter 12: Has What We're Measuring Changed?

1. Ben Quinn, 'Hate crimes double in five years in England and Wales', the
Guardian, 2019 https://www.theguardian.com/society/2019/oct/15/hate-
crimes-double-england-wales
2. Hate Crime statistical bulletin, England and Wales 2018/19, Home Office,
2019 https://assets.publishing.service.gov.uk/government/uploads/system/
uploads/attachment_data/file/839172/hate-crime-1819-hosb2419.pdf
3. Nancy Kelley, Dr Omar Khan and Sarah Sharrock, 'Racial prejudice in
Britain today', NatCen, September 2017 http://natcen.ac.uk/media/1488132/
racial-prejudice-report_v4.pdf
4. Ibid.
5. 'Data & statistics on autism spectrum disorder', US Centers for Disease
Control and Prevention https://www.cdc.gov/ncbddd/autism/data.html
6. Lotter, V., 'Epidemiology of autistic conditions in young children', *Social
Psychiatry*, 1(3) (1966), pp. 124–35.
7. Treffert, D. A., 'Epidemiology of infantile autism', *Archives of General
Psychiatry*, 22(5) (1970), pp. 431–8.
8. Kanner, L., 'Autistic disturbances of affective contact', *Nervous Child*, 2
(1943), pp. 217–50.
9. This explanation is largely taken from Lina Zeldovich, 'The evolution of

"autism" as a diagnosis, explained', *Spectrum News*, 9 May 2018 https://www.spectrumnews.org/news/evolution-autism-diagnosis-explained/

10. Volkmar, F. R., Cohen, D. J. and Paul, R., 'An evaluation of DSM-III criteria for infantile autism', *Journal of the American Academy of Child & Adolescent Psychiatry*, 25(2) (1986), pp. 190–97 doi:10.1016/s0002-7138(09)60226-0

11. 'Crime in England and Wales: Appendix tables', ONS, year ending December 2019 https://www.ons.gov.uk/peoplepopulationandcommunity/crimeandjustice/datasets/crimeinenglandandwalesappendixtables

12. The Law Reports (Appeal Cases), *R* v *R* (1991) UKHL 12 (23 October 1991) http://www.bailii.org/uk/cases/UKHL/1991/12.html

13. 'Sexual offending: Crime Survey for England and Wales appendix tables', ONS, 13 December 2018 https://www.ons.gov.uk/peoplepopulationandcommunity/crimeandjustice/datasets/sexualoffendingcrimesurveyforenglandandwalesappendixtables

14. 'United States: Weekly and biweekly deaths: where are confirmed deaths increasing or falling?', *Our World in Data*, 30 June 2020 update https://ourworldindata.org/coronavirus/country/united-states?country=~USA#weekly-and-biweekly-deaths-where-are-confirmed-deaths-increasing-or-falling

15. House of Commons Library Briefing Paper Number 8537, 2019, Hate Crime Statistics https://commonslibrary.parliament.uk/research-briefings/cbp-8537/

Chapter 13: Rankings

1. Sean Coughlan, 'Pisa tests: UK rises in international school rankings', *BBC News*, 2019 https://www.bbc.co.uk/news/education-50563833

2. 'India surpasses France, UK to become world's 5th largest economy: IMF', *Business Today*, 23 February 2020 https://www.businesstoday.in/current/economy-politics/india-surpasses-france-uk-to-become-world-5th-largest-economy-imf/story/396717.html

3. Alanna Petroff, 'Britain crashes out of world's top 5 economies', *CNN*, 2017 https://money.cnn.com/2017/11/22/news/economy/uk-france-biggest-economies-in-the-world/index.html

4. Darren Boyle, 'India overtakes Britain as the world's sixth largest economy (so why are WE still planning to send THEM £130 million in aid by 2018?)', the *Daily Mail*, 2016 https://www.dailymail.co.uk/news/article-4056296/India-overtakes-Britain-world-s-sixth-largest-economy-earth-planning-send-130-million-aid-end-2018.html

5. World Economic and Financial Surveys, World Economic Outlook Database, IMF.org https://www.imf.org/external/pubs/ft/weo/2019/02/weodata/index.aspx

6. Marcus Stead, 'The quiet death of Virgin Cola', 2012 https://marcussteaduk.wordpress.com/2011/02/20/virgin-cola/

7. Clark, A. E., Frijters, P. and Shields, M. A., 'Relative income, happiness, and Utility: An explanation for the Easterlin Paradox and other puzzles', *Journal of Economic Literature*, 46(1) (2008), pp. 95–144 doi:10.1257/jel.46.1.95

8. IMF World Economic Outlook Database, 2019 https://www.imf.org/external/pubs/ft/weo/2019/02/weodata/index.aspx

9. 'QS World University Rankings: Methodology', 2020 https://www.topuniversities.com/qs-world-university-rankings/methodology

10. 'University league tables 2020', the *Guardian*, https://www.theguardian.com/education/ng-interactive/2019/jun/07/university-league-tables-2020

11. OECD PISA FAQ http://www.oecd.org/pisa/pisafaq/

12. 'PISA 2018 results: Combined executive summaries' https://www.oecd.org/pisa/Combined_Executive_Summaries_PISA_2018.pdf

Chapter 14: Is It Representative of the Literature?

1. Joe Pinkstone, 'Drinking a small glass of red wine a day could help avoid age-related health problems like diabetes, Alzheimer's and heart disease, study finds', the *Daily Mail*, 2020 https://www.dailymail.co.uk/sciencetech/article-8185207/Drinking-small-glass-red-wine-day-good-long-term-health.html

2. Alexandra Thompson, 'A glass of red is NOT good for the heart: Scientists debunk the myth that drinking in moderation has health benefits', the *Daily Mail*, 2017 https://www.dailymail.co.uk/health/article-4529928/A-glass-red-wine-NOT-good-heart.html

3. Alexandra Thompson, 'One glass of antioxidant-rich red wine a day slashes men's risk of prostate cancer by more than 10% – but Chardonnay has the opposite effect, study finds', the *Daily Mail*, 2018 https://www.dailymail.co.uk/health/article-5703883/One-glass-antioxidant-rich-red-wine-day-slashes-mens-risk-prostate-cancer-10.html

4. Colin Fernandez, 'Even one glass of wine a day raises the risk of cancer: Alarming study reveals booze is linked to at least SEVEN forms of the disease', the *Daily Mail*, 2016 https://www.dailymail.co.uk/health/article-3701871/Even-one-glass-wine-day-raises-risk-cancer-Alarming-study-reveals-booze-linked-SEVEN-forms-disease.html

5. Mold, M., Umar, D., King, A. and Exley, C., 'Aluminium in brain tissue in autism', Journal of Trace Elements in Medicine and Biology, 46 (March 2018), pp. 76-82 doi:10.1016/j.jtemb.2017.11.012

6. Chris Exley and Alexandra Thompson, 'Perhaps we now have the link between vaccination and autism: Professor reveals aluminium in jabs may cause sufferers to have up to 10 times more of the metal in their brains

than is safe', the *Daily Mail*, 2017, archived at https://web.archive.org/web/20171130210126/http://www.dailymail.co.uk/health/article-5133049/Aluminium-vaccines-cause-autism.html

7. Wakefield, A. J., Murch, S. H., Anthony, A., Linnell, J. et al., 'RETRACTED: Ileal-lymphoid-nodular hyperplasia, non-specific colitis, and pervasive developmental disorder in children', *The Lancet*, 28 February 1998 https://doi.org/10.1016/S0140-6736(97)11096-0

8. Godlee, F., Smith, J. and Marcovitch, H., 'Wakefield's article linking MMR vaccine and autism was fraudulent', *British Medical Journal*, 342 (2011), c7452.

9. 'More than 140,000 die from measles as cases surge worldwide', WHO, 2019 https://www.who.int/news-room/detail/05-12-2019-more-than-140-000-die-from-measles-as-cases-surge-worldwide

10. Xi, B., Veeranki, S. P., Zhao, M., Ma, C., Yan, Y. and Mi, J., 'Relationship of alcohol consumption to all-cause, cardiovascular, and cancer-related mortality in U.S. adults', *Journal of the American College of Cardiology*, 70(8) (August 2017), pp. 913–22.

11. Bell, S., Daskalopoulou, M., Rapsomaniki, E., George, J., Britton, A., Bobak, M., Casas, J. P., Dale, C. E., Denaxas, S., Shah, A. D. and Hemingway, H., 'Association between clinically recorded alcohol consumption and initial presentation of 12 cardiovascular diseases: Population based cohort study using linked health records', *British Medical Journal*, 356 (2017), j909.

12. Gonçalves, A., Claggett, B., Jhund, P. S., Rosamond, W., Deswal, A., Aguilar, D., Shah, A. M., Cheng, S. and Solomon, S. D., 'Alcohol consumption and risk of heart failure: The atherosclerosis risk in communities study', *European Heart Journal*, 36 (14) (14 April 2015), pp. 939–45 https://doi.org/10.1093/eurheartj/ehu514

Chapter 15: Demand for Novelty

1. Lucy Hooker, 'Does money make you mean?', *BBC News*, 2015 https://www.bbc.co.uk/news/magazine-31761576

2. Vohs, K. D., Mead, N. L. and Goode, M. R., 'The psychological consequences of money', *Science*, 314 (17 November 2006).

3. Daniel Kahneman, *Thinking, Fast and Slow*, Allen Lane, 2011.

4. Bateson, M., Nettle, D. and Roberts, G., 'Cues of being watched enhance cooperation in a real-world setting', *Biology Letters*, 2(3) (2006), pp. 412–14 doi:10.1098/rsbl.2006.0509

5. Zhong, C.-B. and Liljenquist, K., 'Washing away your sins: Threatened morality and physical cleansing', *Science*, 313 (8 September 2006), pp. 1451–2 doi:10.1126/science.1130726.

6. Joe Pinsker, 'Just *looking* at cash makes people selfish and less social', *The*

Atlantic, 2014.

7. Bem, D. J., 'Feeling the future: Experimental evidence for anomalous retroactive influences on cognition and affect', *Journal of Personality and Social Psychology*, 100(3) (2011), pp. 407–25 https://doi.org/10.1037/a0021524

8. Ritchie, S., Wiseman, R. and French, C., 'Replication, replication, replication', *The Psychologist*, 25 (May 2012) https://thepsychologist.bps.org. uk/volume-25/edition-5/replication-replication-replication

9. Ritchie, S. J., Wiseman, R. and French, C. C., 'Failing the future: Three unsuccessful attempts to replicate Bem's "retroactive facilitation of recall" effect', *PLOS One*, 7(3) (2012), e33423 https://doi.org/10.1371/journal. pone.0033423

10. Bem, D., Tressoldi, P. E., Rabeyron, T. and Duggan, M., 'Feeling the future: A meta-analysis of 90 experiments on the anomalous anticipation of random future events (version 2; peer review: 2 approved)', *F1000Research*, 2016, 4:1188 https://doi.org/10.12688/f1000research.7177.2

11. Simes R. J., 'Publication bias: The case for an international registry of clinical trials', *Journal of Clinical Oncology*, 4(10) (1 October 1986), pp. 1529–41 doi:10.1200/JCO.1986.4.10

12. Driessen, E., Hollon, S. D., Bockting, C. L. H., Cuijpers, P. and Turner, E. H., 'Does publication bias inflate the apparent efficacy of psychological treatment for major depressive disorder? A systematic review and meta-analysis of US national institutes of health-funded trials', *PLOS One*, 10(9) (2015), e0137864 doi:10.1371/journal.pone.0137864

13. Conn, V., Valentine, J., Cooper, H. and Rantz, M., 'Grey literature in meta-analyses', *Nursing Research*, 52(4) (2003), pp. 256–61 doi:10.1097/00006199-200307000-00008

14. DeVito, N. J., Bacon, S. and Goldacre, B., 'Compliance with legal requirement to report clinical trial results on ClinicalTrials.gov: A cohort study', *The Lancet*, 17 January 2020 doi:https://doi.org/10.1016/S0140-6736(19)33220-9

15. Lodder, P., Ong, H. H., Grasman, R. P. P. P. and Wicherts, J. M., 'A comprehensive meta-analysis of money priming', *Journal of Experimental Psychology: General*, 148(4) (2019), pp. 688–712 doi:10.1037/xge0000570

16. Scheel, A., Schijen, M. and Lakens, D., 'An excess of positive results: Comparing the standard psychology literature with registered reports', *PsyArVix*, 5 February 2020 doi:10.31234/osf.io/p6e9c

Chapter 16: Cherry-picking

1. Bob Carter, 'There IS a problem with global warming. . . it stopped in 1998,' the *Daily Telegraph*, 2006 https://www.telegraph.co.uk/ comment/3624242/There-IS-a-problem-with-global-warming...-it-stopped-in-1998.html

2. David Rose, 'Global warming stopped 16 years ago, reveals Met Office report quietly released . . . and here is the chart to prove it', the *Mail on Sunday*, 2012 https://www.dailymail.co.uk/sciencetech/article-2217286/ Global-warming-stopped-16-years-ago-reveals-Met-Office-report-quietly-released--chart-prove-it.html

3. Sam Griffiths and Tim Shipman, '"Suicidal generation": tragic toll of teens doubles in 8 years', the *Sunday Times*, 2019 https://www.thetimes. co.uk/edition/news/suicidal-generation-tragic-toll-of-teens-doubles-in-8-years-zlkqzsd2b

4. 'Suicides in the UK: 2018 registrations', ONS, 3 September 2019 https://www.ons.gov.uk/peoplepopulationandcommunity/ birthsdeathsandmarriages/deaths/bulletins/ suicidesintheunitedkingdom/2018registrations

5. COMPare, 'Tracking switched outcomes in medical trials', Centre for Evidence-Based Medicine, 2018 https://compare-trials.org/index.html

Chapter 17: Forecasting

1. Philip Inman, 'OBR caps UK growth forecast at 1.2% but says five-year outlook bright', the *Guardian*, 2019 https://www.theguardian.com/ business/2019/mar/13/obr-caps-uk-growth-forecast-at-12-but-says-five-year-outlook-bright

2. 'How our forecasts measure up', Met Office blog, 2016 https://blog. metoffice.gov.uk/2016/07/10/how-our-forecasts-measure-up/

3. Nate Silver, *The Signal and the Noise: The Art and Science of Prediction*, Penguin 2012.

Chapter 18: Assumptions in Models

1. Peter Hitchens, 'There's powerful evidence this Great Panic is foolish, yet our freedom is still broken and our economy crippled', the *Mail on Sunday*, 2020. Archived at the Wayback Machine: https://www.dailymail.co.uk/ debate/article-8163587/PETER-HITCHENS-Great-Panic-foolish-freedom-broken-economy-crippled.html We have used the archived version because the original *Mail on Sunday* article has been edited since publication, so that the quoted line 'He has twice revised his terrifying prophecy, first to fewer than 20,000 and then on Friday to 5,700', now reads 'He *or others from Imperial College* have twice revised his terrifying prophecy, first to fewer than 20,000 and then on Friday to 5,700' (our italics).

2. 'United Kingdom: What is the cumulative number of confirmed deaths?', *Our World in Data* https://ourworldindata.org/coronavirus/country/ united-kingdom?country=~GBR#what-is-the-cumulative-number-of-confirmed-deaths

3. Ferguson, N. M., Laydon, D., Nedjati-Gilani, G., Imai, N. et al., 'Impact of non-pharmaceutical interventions (NPIs) to reduce COVID-19 mortality and healthcare demand', Imperial College London, 16 March 2020 https://www.imperial.ac.uk/media/imperial-college/medicine/sph/ide/gida-fellowships/Imperial-College-COVID19-NPI-modelling-16-03-2020.pdf

4. Lourenço, J., Paton, R., Ghafari, M., Kraemer, M., Thompson, C., Simmonds, P., Klenerman, P. and Gupta, S., 'Fundamental principles of epidemic spread highlight the immediate need for large-scale serological surveys to assess the stage of the SARS-CoV-2 epidemic', *medRxiv* 2020.03.24.20042291 (preprint) 2020 https://doi.org/10.1101/2020.03.24.20042291

5. Chris Giles, 'Estimates of long-term effect of Brexit on national income' chart, 'Brexit in seven charts – the economic impact', the *Financial Times*, 2016 https://www.ft.com/content/0260242c-370b-11e6-9a05-82a9b15a8ee7

6. 'The economy after Brexit', Economists for Brexit, 2016 http://issuu.com/efbkl/docs/economists_for_brexit_-_the_economy/1?e=24629146/35248609

7. 'HM Treasury analysis: The immediate economic impact of leaving the EU', HM Treasury, 2016 https://assets.publishing.service.gov.uk/government/uploads/system/uploads/attachment_data/file/524967/hm_treasury_analysis_the_immediate_economic_impact_of_leaving_the_eu_web.pdf

8. 'Family spending workbook 1: Detailed expenditure and trends', table 4.3, ONS, 19 March 2020. https://www.ons.gov.uk/peoplepopulationandcommunity/personalandhouseholdfinances/expenditure/datasets/familyspendingworkbook1detailedexpenditureandtrends

9. Estrin, S., Cote, C., and Shapiro, D., 'Can Brexit defy gravity? It is still much cheaper to trade with neighbouring countries', LSE blog, 9 November 2018 https://blogs.lse.ac.uk/management/2018/11/09/can-brexit-defy-gravity-it-is-still-much-cheaper-to-trade-with-neighbouring-countries/

10. Head, K. and Mayer, T., Gravity equations: Workhorse, toolkit, and cookbook', *Handbook of International Economics*, 4(10) (2013) doi:1016/B978-0-444-54314-1.00003-3.

11. Sampson, T., Dhingra, S., Ottaviano, G. and Van Reenan, J., 'Economists for Brexit: A critique', CEP Brexit Analysis Paper No. 6, 2016 http://cep.lse.ac.uk/pubs/download/brexit06.pdf

12. Pike, W. T. and Saini, V., 'An international comparison of the second derivative of COVID-19 deaths after implementation of social distancing measures', *medRxiv* 2020.03.25.20041475 doi:https://doi.org/10.1101/2020.03.25.20041475

13. Tom Pike, Twitter, 2020: https://twitter.com/TomPike00075908/status/1244077827164643328.

Chapter 19: Texas Sharpshooter Fallacy

1. Stefan Shakespeare, 'Introducing YouGov's 2017 election model', YouGov, 2017 https://yougov.co.uk/topics/politics/articles-reports/2017/05/31/yougovs-election-model

2. Stefan Shakespeare, 'How YouGov's election model compares with the final result', YouGov, 2017 https://yougov.co.uk/topics/politics/articles-reports/2017/06/09/how-yougovs-election-model-compares-final-result

3. Anthony Wells, 'Final 2019 general election MRP model: Small Conservative majority likely', YouGov, 2019 https://yougov.co.uk/topics/politics/articles-reports/2019/12/10/final-2019-general-election-mrp-model-small-

4. John Rentoul, 'The new YouGov poll means this election is going to the wire', the *Independent*, 2019 https://www.independent.co.uk/voices/election-yougov-latest-poll-mrp-yougov-survation-tories-labour-majority-hung-parliament-a9241366.html

5. Mia de Graaf, 'Cell phone tower shut down at elementary school after eight kids are diagnosed with cancer in "mysterious" cluster', the *Daily Mail*, 2019 https://www.dailymail.co.uk/health/article-6886561/Cell-phone-tower-shut-elementary-school-eight-kids-diagnosed-cancer.html

6. Julie Watts, 'Moms of kids with cancer turn attention from school cell tower to the water', *CBS Sacramento*, 2019 https://sacramento.cbslocal.com/2019/05/02/moms-kids-cancer-cell-tower-water-ripon/

7. 'Cancer facts & figures 2020', American Cancer Society, Atlanta, Ga., 2020 https://www.cancer.org/research/cancer-facts-statistics/all-cancer-facts-figures/cancer-facts-figures-2020.html

8. Siméon Denis Poisson, *Recherches sur la probabilité des jugements en matière criminelle et en matière civile*, 1837, translated 2013 by Oscar Sheynin https://arxiv.org/pdf/1902.02782.pdf

9. Sam Greenhill, '"It's awful – Why did nobody see it coming?": The Queen gives her verdict on global credit crunch', the *Daily Mail*, 2008 https://www.dailymail.co.uk/news/article-1083290/Its-awful--Why-did-coming--The-Queen-gives-verdict-global-credit-crunch.html

10. House of Commons Hansard Debates, 13 November 2003, Column 397 https://publications.parliament.uk/pa/cm200203/cmhansrd/vo031113/debtext/31113-02.htm

11. Melissa Kite, 'Vince Cable: Sage of the credit crunch, but this Liberal Democrat is not for gloating', the *Daily Telegraph*, 2008 https://www.telegraph.co.uk/news/politics/liberaldemocrats/3179505/Vince-Cable-Sage-of-the-credit-crunch-but-this-Liberal-Democrat-is-not-for-gloating.html

12. Paul Samuelson (1966), quoted in Bluedorn, J. C., Decressin, J. and

Terrones, M. E., 'Do asset price drops foreshadow recessions?', IMF Working Paper, October 2013, p. 4.

13. Asa Bennett, *Romanifesto: Modern Lessons from Classical Politics*, Biteback Publishing, 2019.

14. Rachael Pells, 'British economy "will turn nasty next year", says former Business Secretary Sir Vince Cable', the *Independent*, 2016 https://www.independent.co.uk/news/business/sir-vince-cable-british-economy-will-turn-nasty-next-year-says-man-who-predicted-2008-economic-crash-a7394316.html

15. Feychting, M. and Alhbom, M., 'Magnetic fields and cancer in children residing near Swedish high-voltage power lines', *American Journal of Epidemiology*, 138(7) (1 October 1993), pp. 467–81 https://doi.org/10.1093/oxfordjournals.aje.a116881

16. Andy Coghlan, 'Swedish studies pinpoint power line cancer link', *New Scientist*, 1992 https://www.newscientist.com/article/mg13618450-300-swedish-studies-pinpoint-power-line-cancer-link/

17. Dr John Moulder, 'Electromagnetic fields and human health: Power lines and cancer FAQs', 2007 http://large.stanford.edu/publications/crime/references/moulder/moulder.pdf

18. Richard Gill, 'Lying statistics damn Nurse Lucia de B', 2007 https://www.math.leidenuniv.nl/~gill/lucia.html

19. Ben Goldacre, 'Lucia de Berk – a martyr to stupidity', the *Guardian*, 2010 https://www.badscience.net/2010/04/lucia-de-berk-a-martyr-to-stupidity/

Chapter 20: Survivorship Bias

1. Danuta Kean, 'The Da Vinci Code code: What's the formula for a bestselling book?', the *Guardian*, 2017 https://www.theguardian.com/books/2017/jan/17/the-da-vinci-code-code-whats-the-formula-for-a-bestselling-book

2. Donna Ferguson, 'Want to write a bestselling novel? Use an algorithm', the *Guardian*, 2017 https://www.theguardian.com/money/2017/sep/23/write-bestselling-novel-algorithm-earning-money

3. Hephzibah Anderson, 'The secret code to writing a bestseller', *BBC Culture*, 2016 https://www.bbc.com/culture/article/20160812-the-secret-code-to-writing-a-bestseller

4. Wald, A., 'A method of estimating plane vulnerability based on damage of survivors', Alexandria, Va., Operations Evaluation Group, Center for Naval Analyses, reprint, CRC432, 1980 https://apps.dtic.mil/dtic/tr/fulltext/u2/a091073.pdf

5. Gary Smith, *Standard Deviations: Flawed Assumptions, Tortured Data, and Other Ways to Lie with Statistics*, Overlook Press 2014.

6. Jordan Ellenberg, *How Not to Be Wrong: The Power of Mathematical Thinking*, Penguin Books, 2014, pp. 89–191.

7. Derren Brown, *The System*, Channel 4, 2008 http://derrenbrown.co.uk/shows/the-system/

8. John D. Sutter, 'Norway mass-shooting trial reopens debate on violent video games', CNN, 2012 https://edition.cnn.com/2012/04/19/tech/gaming-gadgets/games-violence-norway-react/index.html

9. Ben Hill, 'From a bullied school boy to NZ's worst mass murderer: Christchurch mosque shooter was "badly picked on as a child because he was chubby"', *Daily Mail Australia*, 2019 https://www.dailymail.co.uk/news/article-6819895/Christchurch-mosque-shooter-picked-pretty-badly-child-overweight.html

10. Jane C. Timm, 'Fact check: Trump suggests video games to blame for mass shootings', NBC News, 2019 https://www.nbcnews.com/politics/donald-trump/fact-check-trump-suggests-video-games-blame-mass-shootings-n1039411

11. Markey, P. M., Markey, C. N. and French, J. E., 'Violent video games and real-world violence: Rhetoric versus data', *Psychology of Popular Media Culture*, 4(4) (2015), pp. 277–95 https://doi.org/10.1037/ppm0000030

12. Turner, E. H., Matthews, A. M., Linardatos, E., Tell, R. A. and Rosenthal, R., 'Selective publication of antidepressant trials and its influence on apparent efficacy', *New England Journal of Medicine*, 358(3) (2008), pp. 252–60 doi:10.1056/NEJMsa065779

Chapter 21: Collider Bias

1. Miyara, M., Tubach, F., Pourcher, V., Morelot-Panzini, C. et al., 'Low incidence of daily active tobacco smoking in patients with symptomatic COVID-19', *Qeios*, 21 April 2020 doi:10.32388/WPP19W.3

2. Mary Kekatos, 'Was Hockney RIGHT? French researchers to give nicotine patches to coronavirus patients and frontline workers after lower rates of infection were found among smokers', the *Daily Mail*, 2020 https://www.dailymail.co.uk/health/article-8246939/French-researchers-plan-nicotine-patches-coronavirus-patients-frontline-workers.html

3. Roberts, R. S., Spitzer, W. O., Delmore, T. and Sackett, D. L., 'An empirical demonstration of Berkson's bias', *Journal of Chronic Diseases*, 31(2) (1978, pp. 119–28 https://doi.org/10.1016/0021-9681(78)90097-8

4. Griffith, G., Morris, T. T., Tudball, M., Herbert, A. et al., 'Collider bias undermines our understanding of COVID-19 disease risk and severity', *medRxiv* 2020.05.04.20090506 doi:https://doi.org/10.1101/2020.05.04.20090506

5. Sperrin, M., Candlish, J., Badrick, E., Renehan, A. and Buchan, I., 'Collider bias is only a partial explanation for the obesity

paradox', *Epidemiology*, 27(4) (July 2016), pp. 525–30 doi:10.1097/
EDE.0000000000000493. PMID: 27075676; PMCID: PMC4890843.

Chapter 22: Goodhart's Law

1. Patrick Worrall, 'The target was for 100,000 tests a day to be "carried
out", not "capacity" to do 100,000 tests', *Channel 4 FactCheck*, 2020 https://
www.channel4.com/news/factcheck/factcheck-the-target-was-for-100000-
tests-a-day-to-be-carried-out-not-capacity-to-do-100000-tests
2. 'United Kingdom: How many tests are performed each day?', *Our
World in Data* https://ourworldindata.org/coronavirus/country/united-
kingdom?country=~GBR#how-many-tests-are-performed-each-day
3. Laura Kuenssberg, Twitter https://twitter.com/bbclaurak/
status/1255757972791230465
4. 'Matt Hancock confirms 100,000 coronavirus testing target met', *ITV
News*, 1 May 2020 https://www.itv.com/news/2020-05-01/coronavirus-daily-
briefing-matt-hancock-steve-powis-testing-tracing/
5. Emily Ashton, Twitter, 29 April 2020 https://twitter.com/elashton/
status/1255468112251695109
6. Nick Carding, 'Government counts mailouts to hit 100,000 testing
target', *Health Service Journal*, 2020 https://www.hsj.co.uk/quality-and-
performance/revealed-how-government-changed-the-rules-to-hit-100000-
tests-target/7027544.article
7. 'More or less: Testing truth, fatality rates obesity risk and trampolines',
BBC Radio 4, 2020 https://www.bbc.co.uk/programmes/p08ccb4g
8. 'Sir David Norgrove response to Matt Hancock regarding the
government's COVID-19 testing data', UK Statistics Authority, 2 June 2020
https://www.statisticsauthority.gov.uk/correspondence/sir-david-norgrove-
response-to-matt-hancock-regarding-the-governments-covid-19-testing-
data/
9. Daisy Christodoulou, 'Exams and Goodhart's law', 2013 https://
daisychristodoulou.com/2013/11/exams-and-goodharts-law/
10. Dave Philipps, 'At veterans hospital in Oregon, a push for better ratings
puts patients at risk, doctors say', the *New York Times*, 2018 https://www.
nytimes.com/2018/01/01/us/at-veterans-hospital-in-oregon-a-push-for-
better-ratings-puts-patients-at-risk-doctors-say.html
11. Gupta, A., Allen, L. A., Bhatt, D. L., Cox, M. et al., 'Association of the
hospital readmissions reduction program implementation with readmission
and mortality outcomes in heart failure', *JAMA Cardiology*, 3(1) (20), pp.
44–53 doi:10.1001/jamacardio.2017.4265
12. Fire, M. and Guestrin, C., 'Over-optimization of academic publishing
metrics: Observing Goodhart's law in action', *GigaScience*, 8(6) (June 2019),
giz053 https://doi.org/10.1093/gigascience/giz053

13. 'Millions more items of PPE for frontline staff from new business partnerships', Gov.uk, 9 May 2020 https://www.gov.uk/government/news/millions-more-items-of-ppe-for-frontline-staff-from-new-business-partnerships

14. Will Kurt, Twitter, 20 May 2016 https://twitter.com/willkurt/status/733708922364657664

15. Tom Chivers, 'Stop obsessing over the 100,000 test target', *UnHerd*, 2020 https://unherd.com/thepost/stop-obsessing-over-the-100000-test-target/

Conclusion and Statistical Style Guide

1. Simon Heffer, *The Daily Telegraph Style Guide*, Aurum Press, 2010.

2. 'Names and titles', Telegraph style book, 23 January 2018 https://www.telegraph.co.uk/style-book/names-and-titles/

3. Emily Favilla, 'BuzzFeed Style Guide', 4 February 2014 https://www.buzzfeed.com/emmyf/buzzfeed-style-guide

4. 'FULL SUNDAY SPORT STYLE GUIDE EMAIL "WHO THE HELL PUTS A HYPHEN IN BELLEND?"', *Guido Fawkes*, 25 July 2014 https://order-order.com/2014/07/25/full-sunday-sport-emailwho-the-hell-puts-a-hyphen-in-bellend/

5. Roger Milne, 'Britain in row with neighbours over North Sea dumping', *New Scientist*, 27 January 1990 https://www.newscientist.com/article/mg12517011-200-britain-in-row-with-neighbours-over-north-sea-dumping/#ixzz6VUDWpXqm

6. Adams, R. C., Challenger, A., Bratton, L., Boivin, J. et al., 'Claims of causality in health news: A randomised trial', *BMC Medicine*, 17, 91 (2019) https://doi.org/10.1186/s12916-019-1324-7

Index

absolute vs relative risk 77–81, 166
aces (cards) 70–1
acting, attractiveness and 151–2
ActiPatch 17, 18
alcohol 97, 100–1
algorithms 123, 141, 147
alternative hypothesis 36–8
Alzheimer's disease, crosswords and 170
Americans, native-born 65–6
anecdotal evidence 15–19
antibody tests 69, 74–5
antidepressants 107, 109, 146–7
apps, weather 119, 120
arthritis, water and 145
assumption, in models 127–32
attractiveness, acting and 151–2
Australia 95, 96
autism 83–6, 100
averages 8–9, 98

back pain, lower 17–19
bacon, cancer risk and 46, 78–9
Baltimore Stockbroker parable 143, 144
Bayes' Theorem 69–75
Bayes, Rev Thomas 70
beer 100
Bem, Daryl 104–6
benefits, and risks 4
bias
    collider 149–55
    publication 98, 106–11, 167–8

sampling 29–34
survivorship 141–7
big numbers 63–7
blood tests 71–2
BMJ (British Medical Journal) 17, 77, 80
bombers, aircraft 141–2
books
    bestselling 141, 147
    reading groups 36–8, 39, 40, 44, 57
    'secrets-of-my-success' 142–3
bowel cancer 78–9
Brexit 130–1
Brier score (forecasting measurement) 121–2
British Social Attitudes survey 83
Brown, Derren 144
burns 15–16

Cable, Vince 137
cancer
    bowel 46, 78–9
    in children 134, 136, 138
    drugs 79
    red wine and 97
    smoking and 56
    studies 107
    terminal 15
cannabis use 47, 50
carbon dioxide emission 47–8, 56
Carter, Bob 113
Cato Institute (think tank) 65–6

causality 55–61, 168
census data 32–3
cherry-picking numbers 4, 113–17,
      168
children
   autism and 83–6, 100
   cancer in 134, 136, 138
   and MMR vaccine 100
   with older fathers 77–8, 80
   school rankings 91
   sleep, and screen time 46
   vaping 47
China 92–3, 95–6, 130–1, 132
Christodoulou, Daisy 159
Clark, Sally 73
climate data 113, 116, 116–17
cola drinks 93
college admissions (US) 152
collider bias 149–55
conditional probability 70–1
confidence interval 168
confounders, controlling for 49–50,
      53, 150, 153
confounding variable 47
consumer spending 128, 137
context, numbers in 166
control groups 57–8
controlled trials 28
Corbyn, Jeremy 30–1
correlation 51–2, 55–6, 58, 60
Countdown 2
counterfactual (control group) 57–8
Covid-19
   controlled trials 28
   deaths 1–2, 2–3
   healthcare workers and 154
   models 125, 127, 132
   R value 7–8, 12–13
   recording deaths (US) 87
   smokers and 149, 154
   tests 154, 157–8, 162
   treatment of 28
crime
   DNA evidence 72–3

figures 65–6
   hate 83, 87–8
   sexual 86–8
Crime Survey for England and
      Wales (CSEW) 86–8
crosswords, Alzheimer's disease
      and 170
cycle journeys 64–5

de Berk, Lucia 138–9
deaths
   Covid-19 1–3
   London cyclists 64–5
   obesity 47–8, 56
   police custody 65
debt, personal 137
Deen, Andrew 72, 73
Denmark 96
denominators 63–7
dependent variable 145, 150–3
diabetes, obesity and 155
Diagnostic and Statistical Manual
      of Mental Disorders editions
      84–5
dice rolling 24–5, 120, 136, 144–5
directed acyclic graph 150, 153
distribution, normal (graph) 23, 98
distribution of events 16–17
DNA evidence 72
drownings, ice cream and 48
drugs
   antidepressants 107, 109, 146–7
   back pain 17
   polls 33
   relative effect measures 79
   studies 26–7

ecological fallacy 13
economic gravity 130–1
economic growth
   Cable on 137
   forecasting 119, 123, 126
   UK world ranking 91–2
   war and 60–1

Economists for Brexit 128, 130–1
education
    GCSE grades 158–9, 161
    gender differences 11–12
    levels 9–12
    school rankings 91
effect size 43–6, 47
elections
    UK 30–1, 33, 133, 139, 170
    US 32
Elizabeth II, Queen 137
Ellenberg, Jordan 143, 144
empirical evidence 25, 125, 129
errors 51, 121–2, 169
EU referendum 128

Facebook 9, 147
false positive/negative 39, 40, 72–3,
    74
fast-growing, meaning of 79
fathers, older 77–8, 80
FDA (US Food and Drug Adminis-
    tration) 18, 69, 79
Ferguson, Prof Neil 127, 132
financial crisis (2008) 128, 129, 137
fish fingers 98–9
fizzy drinks 55, 93
food poisoning, and influenza 153
football rankings 92
football surveys 30
forecasting 119–26, 168
France 91–4, 130–1
French, Chris 106
funnel plots 107–9

Gallup 32
GDP (gross domestic product) 48,
    92–4, 126, 128
gender differences
    eating to impress women 35,
        40–1
    grad school 11–12
    watching Grey's Anatomy 32–3
Germany 92–3, 96

Gill, Richard 138–9
global warming 113, 116–17
Goldacre, Ben 138–9
Goodhart, Charles 158
Goodhart's Law 157–62
gravity, economic 130–1
gravity, physical 16, 130
grey hair, running and 149–50
Grey's Anatomy 32–3

Hancock, Matt 157–8
HARKing (hypothesising after
    results are known) 115–16
hate crime 83, 87–8
Hawthorne effect 57
healthcare 154, 159–60
height, measuring 21–3, 26, 29,
    50–3
Hitchens, Peter 127, 131–2
Home Office 65
homeopathic remedies 18
homicides 65–6
Hospital Readmissions Reduction
    Program (HRRP) 160
house styles 163–5
human behaviour, predicting 123,
    132
humming 109–10
hypothesis testing (significance
    testing) 36–8

ice cream, and drownings 48
images, subliminal 105
IMF (International Monetary Fund)
    91–2, 94
immigrants, undocumented 65–6
immunity passports 69, 73–4
Imperial College London 127, 131–2
independent variables 150–3
India 91–3
individual studies 98, 105, 167
Indonesia 94
infectious-disease models 125, 130
influenza, food poisoning ad 153

instrumental variable approach
        60–1
internet polls 32
investment funds 143–4

J-shaped curve 101
Japan 92, 95–6
Johnson, Boris 30–1

Kahneman, Daniel 103
Kaufman, Amie 1
King's College London 95
Kurt, Will 162

'La Marseillaise' (French anthem)
        109–10
least squares method 51–2
Liechtenstein 94
lifetime earnings, military and 58
lighting, production and 57
line of best fit 52
Literary Digest magazine 32
literature, representation of 97–101,
        166–7
lockdown 29, 69, 119, 127, 132
LSE (London School of Economics)
        131
lung cancer 56

Manchester, University of 95
marijuana use 47, 49–50
mass shootings 145
maths, understanding 2
mean average 8–9
measles 100
measurement practices, changing
        83–9
median average 8–9, 12
median wage (US) 8–10
medical testing 71–2
medicine 15–19, 33
mental arithmetic 2
meta-analyses 47, 49, 72, 106
metrics, using 158–62

Mexico 15, 96
military, and lifetime earnings 58
Minford, Patrick 128
mislead, how numbers can 7–13
MMR vaccine 100
mobile-phone masts (US) 134
mode average 8–9
models, assumption in 127–32
money priming 103–5, 110
mortality rates 101, 160
Mosteller, Frederick 7
multilevel regression with post-
        stratification (MRP) model
        133, 139
murders 65–6, 73, 138–9, 145–6
mutual exclusivity 136

natural experiments 58
Navy, US 141–2
negative rate (specificity) 71
New Zealand 96
NHS, funding 63, 66
Norway 96
novelty, demand for 103–11
null hypothesis 36–8
numerators 63

obesity 47–8, 56, 155
observational studies 56–61
Office for Budget Responsibility
        (OBR) 119, 126

p-hacking 41–2, 98, 110, 167–8
p-value (probability value) 38, 41, 97
pavements, wet 56
pen names, androgynous 141, 147
people, as numbers 1–2
pervasive developmental disorder
        – not otherwise specified'
        (PDD-NOS) 85
Philippines 96
PISA rankings (education) 91, 95–6
Poisson distribution formula 134–6
Poisson, Siméon Denis 135

polls, election 30–1, 33, 124, 133
positive association (positive correlation) 51
positive rate (sensitivity) 71
power lines 138
power, statistical 27
predictions 119–25, 131, 137
priming, money 103–5, 110
priming, social 103–5
prior probability 70–4
probability
    dice rolling 24–5
    prior 70–4
    value (p-value) 38, 41, 97
prosecutor's fallacy 72–3
psychic ability 104–7
publication bias 98, 106–11, 167–8
publish or perish model (academia) 110, 160

quality-of-care ratings 159–60

R value (reproductive number) 7–8, 12–13
rain 56, 119–25
Raisin.co.uk 29, 34
randomised controlled trial (RCT) 57–9
randomness 26, 32, 38, 98, 139, 168
rankings 91–6, 169
rape 72, 86
Registered Reports (RRs) 111
regression, statistical 50–3
relative effect measures 79
relative vs absolute risk 77–81, 166
replication crisis 41–2, 104
representation, of literature 97–101, 166–7
residual (error) 51
risks, benefits and 4
risky behaviour 49
Ritchie, Stuart 106
running, and grey hair 149–50

sample sizes 21–34, 44, 97–8, 167
sampling bias 29–34
Samuelson, Paul 137
school rankings 91
screen time, sleep and 43–4, 45–6
sensation-seeking 49–50
sensitivity (positive rate) 71
sexual crime 86–8
significance, statistical 35–42, 43–5
significance testing (hypothesis testing) 36–8
Simes, R. J. 107
Simpson, Edward H. 10
Simpson's paradox 10–12
SIR model (infectious disease) 125
sleep, and screen time 43–4, 45–6
Sliding Doors (film) 37
Smith, Gary 143
smoking
    alternatives to 47
    Covid-19 and 149, 154
    education levels and 11
    and lung cancer 56
snacks, popular 29
snoring 98–9
social attitudes 83
social media 9, 30–1, 144, 147
social priming 103–5
sources, importance of 169
specificity (negative rate) 71
spending, UK government 63, 66
squared error 121–2
strength experiments 27
style guides 163–71
subliminal advertising 105
Sudden Infant Death Syndrome (SIDS) 73
suicide rates 115
survivorship bias 141–7
swearing study 21, 27–8
Sweden 96, 138

Taiwan 96

teenagers
    fizzy drinks and 55
    risky behaviour 49
    sleep and screen time 46
    suicide rates 115
tests, statistical 36–8, 57
Texas Hold'em poker 120
Texas sharpshooter fallacy 136–9
Texas, USA 65–6
text-mining soft ware 141
therapies, alternative 15, 17–19
toast, burnt 147
trade, international 130–1
Transport for London 64
Treasury 128–9
Trump, Donald 28, 65, 145
Tversky, Amos 103
Twitter 30–1, 144

UK Statistics Authority 63
uncertainty interval 123, 126, 168
unemployment 119, 128
United Kingdom (UK) 91–4, 95–6
United States (US) 92–3, 96, 134
user engagement, media 160

vaccination scares 84, 100

vaping 47, 49–50
variance
    dependent variables 150–3
    height 23–4
    independent variables 150–3
    random 168–9
video games 145–6, 147
Vietnam War 58
Virgin Cola 93

wage, median (US) 8–10, 12
Wakefield, Andrew 100
Wald, Abraham 142
Wansink, Brian 35, 40–2
water and arthritis 145
waves 113–15
weather forecasting 119–25, 129–30
weight, measuring 50–3
Wicks, Joe 26
wine, red 97, 100
Wiseman, Richard 106
World University Rankings score
    94–5

YouGov 30–1, 133, 139